Complete Science Communication
A Guide to Connecting with Scientists, Jo

Complete Science Communication
A Guide to Connecting with Scientists, Journalists and the Public

Ryan C. Fortenberry
University of Mississippi, USA
Email: r410@olemiss.edu

Print ISBN: 978-1-78801-110-5
EPUB ISBN: 978-1-78801-548-6

A catalogue record for this book is available from the British Library

The Royal Society of Chemistry is a charity, registered in England and Wales, Number 207890, and a company incorporated in England by Royal Charter (Registered No. RC000524), registered office: Burlington House, Piccadilly, London W1J 0BA, UK, Telephone: +44 (0) 20 7437 8656.

Visit our website at www.rsc.org/books

Printed in the United Kingdom by CPI Group (UK) Ltd, Croydon, CR0 4YY, UK

Preface

This book arose out of necessity. When I arrived at Georgia Southern University (GSU), the new Master's of Science in Applied Physical Science (MS-APS) was in its inaugural year. The Professional Science Master's option had been approved by the University System of Georgia Board of Regents in no small part due to its bridging of science with business. Part of the curriculum was a writing course. That year some students took a course from another department in the university to satisfy this requirement, and I asked how it went. They all hated the course. Yes, students are supposed to hate courses, but I also asked them about if they had learned how to give a talk, write a paper, or make a sales pitch. Suffice it to say, they had not.

In an earlier life, I had wanted to be a science journalist. I earned a Master's of Science in Communication from Mississippi College and enrolled at Virginia Tech to pursue a Ph.D. in quantum chemistry. I decided that any good science writer should also be a scientist, and I had done applied mathematics research as an undergraduate. I got to Virginia Tech and found that this stuff was hard. I hated it. However, my advisor had the foresight to ask me what I really liked. The answer was astronomy, and he introduced me to astrochemistry. I fell in love. The rest is history.

All of this to say that my background in communication mixed with science was always a passion. I had taught a communication course in writing about science for many semesters as an online adjunct at Mississippi College, and what I heard my students experiencing about

Complete Science Communication: A Guide to Connecting with Scientists, Journalists and the Public
By Ryan C. Fortenberry
© Ryan C. Fortenberry 2019
Published by the Royal Society of Chemistry, www.rsc.org

writing in the GSU MS-APS course made me want to scream. I campaigned and volunteered myself to teach the GSU science communication course. I was stoked. Finally, I would fix this for our MS-APS students. I only needed to find a text book.

I searched high and low across publishers and even disciplines, but nothing fit what I felt was necessary. They were all campy back-slapping encouragement affairs, narrowly targeted at writing only proposals, or the like. As a result, I punted the textbook and simply decided to teach my class the way I thought it needed to be taught: write like a journalist; speak like a caveman. I borrowed David Hilbert's technique of letting the class notes be the textbook. What could go wrong?

Hubris is an amazing gut-punch, and I was waiting for push-back from my colleagues when my students started turning in abstracts and papers to their research mentors and giving presentations for the department. To my surprise, my colleagues loved the way the students were writing and speaking. I thought I was going to have to fight these folks to overcome their bias from their own graduate school instruction. Instead, most of them saw the benefits of what the students were doing and embraced this writing and speaking mindset fully.

I was approached the following summer after I had taught this class for the first time to write a book on astrochemistry. I was organizing a symposium at the American Chemical Society meeting on the subject, and the Royal Society of Chemistry wanted to explore this with me. There are lots of really good astrochemistry books and texts by folks more qualified than me. I politely declined, but, on a whim, threw out to the editor, "You know what would really be useful? A science communication book like nothing you've seen before." The editor was intrigued and asked if I would be willing to write a proposal. I did, eventually.

I was not too enthused by the idea, to be honest. Science communication as a subject lacked for me the sexiness of astrochemistry at the time, and writing a book is daunting. However, my father and my wife convinced me that this played to my strengths and experiences. As a result, I wrote a proposal for a book at which I thought everyone would scoff. Who would want to promote the communication of science as writing like a journalist and speaking like a caveman? To my surprise, it got picked up. I wrote it while teaching the GSU MS-APS science communication course a second time giving me both internal and external processing of ideas. Plus, the students helped me to clarify and streamline my points as I instructed them. I am pleased with how this turned out, and I am fairly confident that no book like this exists on the planet, yet. As I transition to the University of Mississippi in my home state, I hope to continue to inform students about the best practices of communication.

This book can be utilized as a textbook but can also be for one's own self-education even up to senior scientists. This text is targeted at scientists to learn communication not journalists who wish to learn science communication. The book should be approachable for undergraduate students provided that they have some experience with the research process. Experience in a research methods, analytical chemistry, or similar course should be sufficient. Even so, senior-level undergraduates and graduate students at any level would benefit the most from these communication considerations put forth herein.

These tenets work to make communication of science from the actual scientist more effective. They also make it easier, more natural, and more enjoyable. Try it and see for yourself.

Ryan C. Fortenberry

Acknowledgements

To begin saying "Thank you" is an inevitable process of leaving some-one out. Hence, if you feel like you were part of my journey here, you were. To you I say, "Thank you."

Specifically, I must first thank my father who was both personally and professionally my communication mentor. I must also thank my mother who kept me focused on my goals and dreams. This time was full of ups and downs for me and my wife, the real writer, but she encouraged me even when her dreams had to hit pause. I must also thank my children for inspiring me to future days. My extended family, both living and passed, who created both my genes and my experiences must be acknowledged for their support, as well.

I must also thank:

- My editors at the Royal Society of Chemistry: Connor Sheppard, Leanne Marle, Lindsay Stewart, and Robin Driscoll.
- My GSU professional science communication (CHEM 6030) stu-dents from both 2016 and 2018: Ian Byrd, Walter Jackson, Richard Govan, Matthew Bassett, Xiaomeng Wang, and Simpo Rose Ogwa Akech; Souleymane Seye, Oladayo Ariyo, Dominique Kornegay, Aaron Quarterman, Alexis Burdette, Kayla Anderson, Unodinma Ofulue, Ifeoluwa Olabampe, Obehi Akhibi, Shelby Scherer, and Aimee Lorts.
- My MC Communication Department teachers: Merle Ziegler, Web Drake, Billy Lytal, Jimmy Hutto, Tim Nicholas, Vicki Williams, and Mignon Kucia.

Complete Science Communication: A Guide to Connecting with Scientists, Journalists and the Public
By Ryan C. Fortenberry
© Ryan C. Fortenberry 2019
Published by the Royal Society of Chemistry, www.rsc.org

- My research mentors: David H. Magers, Tommy Leavelle, T. Daniel Crawford, and Timothy J. Lee.
- My collaborators and professional friends: Joseph S. Francisco, Susanna Widicus Weaver, Xinchuan Huang, Taylor J. Mach, Joshua P. Layfield, D. Brandon Magers, Ashley Ringer McDonald, David Sherrill, Tim Fuhrer, Callie Cole, Partha Bera, Ella Sciamm-O'Brien, Naseem Rangwala, Michel Nuevo, Florian Bischoff, David Woon, Steven R. Gwaltney, Russell Thackston, Bill Peters, Nathan J. DeYonker, Michael A. Duncan, Michael C. McCarthy, Joel M. Bowman, and Xander Tielens.
- My GSU Chemistry colleagues, specifically: Mike Hurst, Nathan Takas, Amanda M. Stewart, Jim LoBue, Allison Amonette, Michele McGibony, John DiCesare, Karelle Aiken, Ria Ramotaur, John Stone, Shainaz Landge, Eric Gato, Eric Johnson, Brian Koehler, Beulah Narendrapurapu, Jeff and Jess Orvis, Rafael Quirino, and Christine Whitlock.
- My research students: Atsu Agbaglo, Quynh Nguyen, Zach Palmer, Cody Stephan, Carlie Novak, Katie Kloska, Dinahlee Lemaistre, Matt Bassett, Owen Hurst, Shelby Scherer, Riley Theis, Jerry Filipek, W. James Morgan, Mallory Theis Green, Megan Moore, Joseph Lukemire, Zach Lee, Mason Kitchens, Erin Seal, Michelle Berlyoung, and Jordan Enyard.
- The kind reviewers of this book who volunteered their time to make sure that it is worth reading.
- ACS Astrochemistry.
- Colby, Lorri, Mike, Owen, and Willow Hurst for being my other family when I needed you.

Finally, thank you for reading.

Ryan C. Fortenberry
University of Mississippi, Department of Chemistry
and Biochemistry, University, MS 38677 USA.
E-mail: r410@olemiss.edu

Contents

Complete Science Communication: A Guide to Connecting with Scientists, Journalists and the Public
By Ryan C. Fortenberry
© Ryan C. Fortenberry 2019
Published by the Royal Society of Chemistry, www.rsc.org

1 The Art and Motivation of Science Communication

1.1 Introduction

The most important sentence in an entire document, any document, is the very first sentence. If a reader makes it past the title, the only thing that he or she is guaranteed to read is the first sentence of the first section. "This must be distinctly understood, or nothing wonderful can come of the story I am going to relate," to quote Dickens. The first sentence is everything. Then, if luck should have it and the reader continues, the rest of the first paragraph will elaborate on this one idea to add a little context and further information. The rest of the document supports the first paragraph which, in turn, supports the first sentence. This present text follows that same format. It would be hypocritical for it not to do so. Hopefully, that has already been guessed.

While every reader subconsciously follows this tenet, as writers we often neglect this prime example of our Pleistocene brains. If something is not immediately beneficial or dangerous, we move on; we forget it. It is not useful. There is another berry to gather, another predator to detect, another rival tribesman to scare away. As advanced as we are, largely due to science, our brains still function in this tribal, hunter-gatherer fashion. While some view this as a limitation on humanity to be changed, in truth, it is what makes us human and can often empower our dazzling insights. Hence, as scientists and communicators, we must understand this limitation and know how to employ it to our advantage and not fight it.

Complete Science Communication: A Guide to Connecting with Scientists, Journalists and the Public
By Ryan C. Fortenberry
© Ryan C. Fortenberry 2019
Published by the Royal Society of Chemistry, www.rsc.org

1.2   The Written Word

If science is not written down, it may as well have never been done. The transfer of information from one person to the next is the key to furthering knowledge such that each individual does not (nearly literally) have to reinvent the wheel and build the same knowledge from nothing. The collapse of civilization only takes place when the lessons of the past are not learned. Science must be written down so that its lessons can be learned whether in success or failure, so that this information can be transferred from one human brain to another. Repeating things over and over again expecting a different result was Einstein's definition of insanity, but how can one know if something has already been done if it is not recorded for posterity through an accessible medium in some timeless fashion? The most timeless medium is the written word.

Chimpanzees (with their 98% genetic similarity to humans) provide their young with nurturing and life skills passed down from one generation to the next. Some have found novel ways of breaking sticks to dig out insects from burrows. Others know that some trees are better for climbing than others. Still more are experts in getting delicious snacks like honey through the use of specialized tree-burrowing techniques they have devised. However, if the family of chimpanzees is wiped out except for a single infant, that lone survivor will never have the knowledge of his ancestors. In theory, if all the chimps could tell the other chimps about their advances, they would be immensely more capable for having that knowledge. This is truly what separates humans from primates, and that pursuit of knowledge through experiment is science. Chimps, dolphins, and even bird-brains like turkeys can communicate distinctively with one another, but like all other species cannot access this exact information again once auditory communication ceases.

The written word is not natural for humans, either. Like chimpanzees, we are social creatures knit together by shared experiences rehashed in story form dozens of times. However, our ability to process symbols and meaning from otherwise useless objects such as religious trinkets, jewelry, and icons paved the way for the advancement of permanently stored information. Initially, civilization and its rise dictated that numerical records be kept, likely the earliest form of written text. Hence, instead of one manager or a handful of record keepers having to memorize the entire finances of a kingdom, written text offered the ability to keep track, make copies, and reduce mistakes. As a result, the human brain's amazing ability to store vast quantities of information could be employed in more advanced ways.

But how do these symbols we associated with numbers or words actually work? This is rarely considered by most of us who take it for granted that a "C" conjoined to an "H" makes the same sound that begins the word associated with common poultry. We learn that this is the sound for the symbol and simply move on. In truth, there is not much more to it than that. Symbols are chosen to represent things that are spoken whether through parts of syllables (Latin and Germanic scripts), whole syllables (Cherokee script), or whole words (Chinese and most other East Asian scripts). However, the rub comes in understanding that symbols are not just written words but spoken words.

Our brains perceive the world in ways as unique to each of us as our fingerprints, retinal patterns, and DNA. We then encode this information into words. Sometimes these are spoken. Sometimes, with modern humans, they are written. In either case, they must be decoded by a receiver such that his or her brain can interpret these symbols based on his or her own experiences. The basic idea for this process is called the Shannon–Weaver model of communication from a 1948 publication by these two individuals entitled *A Mathematical Theory of Communication* and is depicted in Figure 1.1. A sender encodes

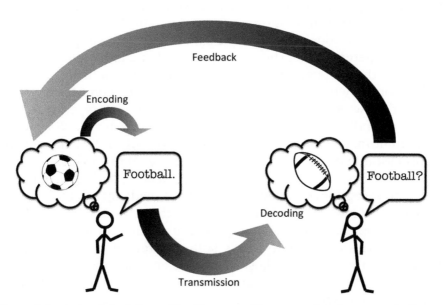

Figure 1.1 A visual depiction of the Shannon–Weaver model of communication.

a message, and transmits it through a medium to be acquired and decoded by a receiver. This encoding is the creation of symbols. Standardization of such symbols is the essence of language. What these symbols actually are is purely a choice of common ground between sender and receiver.

This idea of common ground is essential for any form of communication. The two parties must have enough in common (usually language but even beyond that in shared experiences) in order for the concepts of one to be understood by the other. Forming common ground with any receiver is the key to effective communication. Note from Figure 1.1 how seemingly simple symbols can be easily misinterpreted. This is why standardization of symbols is necessary for communication and even what so often defines culture.

Most of our words have derived from a need to survive, whether through the procurement of food or to fend off invaders or predators, but all stem from a need to communicate with other people to accomplish tasks that we cannot (or choose not) to perform on our own. As a result, different cultures have different words for the same thing. Sometimes one word in one language can have several subtypes of the item being described and encoded to delineate. Snow is the classic example with its one English word and dozens in languages of the First Nations Peoples of Alaska and Canada. Different cultures have different reasons for encoding different things. The San or Basarwa people of the Kalahari Desert (as well as the Xhosa of South Africa) even have more consonants than European languages do. They add various clicks and chirps. The reason is simple. As a true hunter-gatherer people, the fewer sounds that can be made to communicate the same amount of information, the less likely game are to be spooked when members of the hunting party converse with one another. However, it is within these confines of history, culture, survival, art, and even politics that languages have evolved. Now, we as modern humans have to take this hodgepodge of context and encode information that such language was never meant to encode.

We invent our own words for new experiences, and science is certainly no exception. Granted, most of these made-up scientific words have roots in modern or arcane languages that are often broken down to their simplest forms and joined back in a different way. For example, the series of bones up one's spine, vertebrae, comes from the Latin for "to turn" which those of us who enjoy yoga can appreciate. However, this word has now also been co-opted by mathematicians, engineers, and scientists to refer to the direction perpendicular to the horizon, where the latter word has its own etymology. Our spines go up and down in the "vertical" direction. "Vertical" now has nothing to do with turning, but it has everything to do with how your backbone rises away from the ground. As a result, these words frequently become specialized, specific, and esoteric. In many ways this is a good thing. Just like the San of Botswana, communicating the same amount of information in as few symbols as possible is a very good idea. This reduces error and communication time such that time-sensitive information can be discussed quickly. However, it can be a nightmare to learn because the new usage and original meaning diverge so strongly that common ground is impossible to establish. As such, science (or any technical level of communication) fits many definitions of a new language replete with its own dialects such as chemistry, physics, biology, and the like.

Reading new words again and again in context is almost no different to our brains than hearing them over and over in conversation. After all, they are simply symbols for our brain to decode, whether heard or read, and put into the context of our personal experiences and memories. As a result, writing science is the best way to keep it for posterity and to enable others to learn this new language. Luckily, most of the syntax and grammar of "the language of science" is the same as the parent tongue. Furthermore, we can return to the same information repeatedly and cogitate on the implication of those symbols potentially in new ways each time; Buddhist monks have been doing this for millennia and finding new meaning in old texts.

The printing press, the age of enlightenment, and the age of exploration all largely coincided (largely not by coincidence) leading Europeans to look beyond their typical experiences and expand their knowledge. Sailors now could consult maps with numerical and textual information to point them in directions necessary to grow trade, which also benefitted from being recorded. The quartermaster or navigator could easily be injured or killed, rendering the mission doomed in one way or another were it not for permanent records. The pilots and captains could also make comments *in situ* in the ship's log in order to share later with shipbuilders about what worked well in watercraft design and what did not. These bits of information could be shared across crews and nations, and such maps or ship plans were highly prized booty for privateers or pirates.

Beyond these primal and early technology examples, writing information is the easiest way to share data without losing it. In the modern context, data are written and shared electronically instead of in print. They are no less useful; probably more so, in fact. Data and results can be shared instantaneously across the globe bringing together practitioners and experts who otherwise would have no other means of connecting. Newton may have never left England thanks to the postal service, and today he would have even less need to do so. As a result, the text and information driven by electronic media make it necessary to take the hunter-gatherer words and contexts and make an art form of them, to communicate ideas that they were never intended to communicate, but must. The art comes in taking one thing and creating something new from it, often in ways never intended. Such is the art of science communication. In this way, scientific ideas will continue to double our knowledge every few months, and the information will be accessible to those who need or even want to know it.

In this text, popular and scientific writing will each be covered as whole chapters. Writing is the anchor of the scientific discourse as most scientific arguments take place through research papers and not in personal interactions (although such have happened; scientists are people after all). All scientific writing must be clear, concise, and correct, but it must also be tailored for the audience. We do invent special words that not all laymen know, after all. This text will teach new principles of writing science as an art form in order to make it the most effective it can be, even with our hunter-gatherer, Shannon–Weaver, symbolic vernacular.

1.3   Communication

If science is not shared with others, it might as well have never been done. This is true whether the science is encoded in spoken or written symbols, or for interpretation by experts or non-experts. Two individuals look at the same problem differently. But by increasing the pool of potential problem solvers to all of humanity, the odds of difficult problems being solved collaboratively increases greatly. Additionally, if advancements that lead to longer, happier, healthier lives are made, non-experts must be told so that they can enjoy the benefits of someone else's scientific advancement. Science must be shared.

The Wright brothers are credited with inventing the motorized, heavier-than-air airplane and successfully flying it in late 1903. However, they were so concerned that their idea would be stolen that few people actually heard about the initial event until 1904. Some in Europe even publicly denounced them as frauds. However, their tours of the United States and Europe eventually convinced most that they had done what they said, largely due to the skill in controlling the aircraft that no novice could exhibit. Had their idea and successful action not garnered such subsequent acceptance, the history of flight could have been marked in very different ways, and the license plates for the state of North Carolina would have needed a different motto from "First in Flight." It all came down to communication. The idea and breakthrough was not initially communicated, making many doubt its validity.

The basic idea behind communication is dissemination, the ability to spread an idea through a group of people. This dissemination is key to the advancement of science. The mass media were first established in the Anglophone world with the printed word in 1476 by William Caxton as the first English language printer. He printed books, but later his output grew to include pamphlets and newspapers. Arguably, one of the largest success of the printed word was the American Revolution. The ideas of liberty and self-determination away from the aloof Parliament in London were printed to stir the hearts of the Colonials to the point where they were willing to sacrifice themselves for ideas of people whom they may have never met. Ideas were passed from one individual to another, and a nation was born. The audience of the masses was eventually also reached through radio, television, and now the Internet, the unifier of all media.

The printing press along with radio and television limited the power to send out messages to a select few who had the funds to initiate any of these enterprises. Those who wished to submit messages to the masses had to court those who had control or access. This led to a reasonable amount of trust in those who controlled these outlets and a marked amount of prestige. The Internet has changed that. Anyone can voice anything. We see this in the way that people choose and select information that they wish to hear or not to hear, leading to warped and misinformed opinions. However, the role of those with access has not changed in some spheres. In science we call this peer-review.

The goal of peer-review is to provide a means of self-regulation for the quality of information that is disseminated. Once an idea is published, especially today on the Internet, it is impossible for it to be retracted and deleted. Andrew Wakefield's infamous autism debacle is a classic example. In any case, peer-review is still alive and well in science, as it should be even with a few erroneous pieces making it through. Those sources that still hold rigorous peer-review are still revered, and, to their credit, those of the highest respect have been able to adapt to the non-printed, electronic word.

While publishers and houses of peer-review are a natural source for the expert or curious albeit informed bystander, the layperson likely gets his or her information from less reputable but no less convincing sources. As science communicators, we must learn how to harness this untamed beast and communicate with individuals again through their Pleistocene brains and Middle Age languages. Journalists have been doing this for centuries. It is time that we as scientists learned the time-tested skills of journalism in order to effectively disseminate our knowledge. Later chapters of this text also argue that such journalistic techniques should even be brought into the hallowed halls of peer-review. As culture and society change so does language. This will creep into true scientific writing, as well, and we, as scientists, must not fight this tide but learn to employ the wave.

1.4   Deciphering Science/Choosing an Audience

If science is not interpretable, it might as well have never been done. In truth, the audience is what chooses the communication. As mentioned above, in science words are invented to communicate new ideas, but these can be foreign to those not trained in the same field. The goal of science communication is to make sure that new ideas are made available to those who "need to know." This is not about keeping national secrets classified, but about enabling the proper people to hear the proper information in as clear, concise, and correct means as possible.

If an asteroid is making its way towards the Earth, astronomers and planetary scientists would need to discuss the impending tragedy in a means different from engineers and soldiers. Politicians and policymakers would need to discuss this same topic very differently as well, and the people of this planet would need certain information in certain ways to ensure peace until the first two groups (scientists and engineers) could figure out a solution. All parties would need to know the information, but what that information might be and how they should receive it must be tailored uniquely for each audience. The wrong information to the wrong group could be inadequate or may incite bedlam. The audience will determine how a message is encoded. Without foreknowledge of the audience, the science communication is doomed to either insult the audience's intelligence or create a message that cannot be grasped. Even writing technically for peer-review has its own challenges. A research paper written for *Lab on a Chip* will utilize different language than the same paper submitted to *Chemical Science* which would be different from the same article destined for *Nature* or even *Chemistry World* in the extreme example where all are still designed for chemists or scientists. The level of expertise in each reader is different. Consequently, the encoding of what is, for all intents and purposes, the same message but with different symbols, is chosen based upon common ground between the author and the audience.

This choice of wording of the same message for different audiences is a natural process for humans. Most commonly, "code-switching" is the term applied to this behavior. We engage in this all of the time. This is easily recognizable with our children. We tell them things in ways that we would not with our adult friends. In fact, the common retort idiom for being spoken down to is, "Don't speak to me like a child." The audience drives the context, the language, and, for science communicators, ultimately the venue or medium we choose to utilize in order to disseminate our message. The late media philosopher Marshall McLuhan succinctly used the phrase, "The medium is the message" to communicate this idea. How and through what means we choose to say something is often just as important as what we say.

As a result, scientists are natural code-switchers. We talk to our non-scientist family members about our work regularly, but differently than we do with our colleagues and collaborators. However, we need to take those lessons and apply them even to technical and peer-reviewed work. Business and law learned the lessons of journalism long ago. The *Wall Street Journal*, nominally a regional business trade newspaper, is now one of the most trusted sources of information on the planet. Hence, science and its advances need to overcome the ivory tower dialects and embrace the natural human means of communication. If we can do this, science can penetrate society to a greater extent while still not sacrificing its chief tenets. Then, the audiences will still flow, but the messages will be more easily broadcast to those who need to know.

For scientists, the easiest audience on whom to try new communication strategies and novel ideas on message encoding is other scientists. They will forgive us if we fail to communicate in anything but technical terms. However, this leads to a closed loop and further esoterica. The layperson should always be our ultimate audience. If one can communicate to a non-expert, surely one can communicate with an expert. Yes, the audience still drives the message, but we should not stray too far away from the journalistic model in order to properly make this art form more useful. As such, this text gets out of the scientific comfort zone that most of us have built for ourselves and goes straight to science as journalism. Let us begin this discussion there, in the precept of journalism as applied to science.

1.5   Assignments

The course described in this book is typically organized into three major sections: Journalism, Technical Writing, and Public Speaking. Within each section, there are several assignments that can be given on a weekly or daily basis with one major assignment serving as a culmination for the entire section. At the end, there is a final Public Relations section and assignment that serves to synthesize all of these ideas together. Typically, the weekly assignments are all averaged together for 30% of the final grade, the three major assignments are another 30% of the final grade, the final Public Relations project is another 30% of the final grade, and 10% of the grade is left for class participation. Unless noted as "major" or "final," all assignments given in the final sections of each chapter should be interpreted to be weekly or daily assignments.

1.5.1   Science in Entertainment Media

This week the students will be introduced to both the nature of journalism and the nature of science communication. This begins the first section (of three) in the course where the theory and concepts of science communication are developed.

   Assignment for Students: Write and upload a full two-page report in Royal Society of Chemistry style (you will need a cover page, the text, and references if you have them) on one of the following programs, answering and discussing the specified questions. Be sure you clearly state to which program(s) your paper corresponds. Some assignments involve programs where the content may be deemed inappropriate. If you feel such about a given program, simply choose another.

- Watch at least two episodes of the American CBS sitcom *The Big Bang Theory* from any season. State the episode titles, and respond to these questions: What actually is the story: the science or the character's lives? Is the science necessary for the plot? To what scientific principles were you newly exposed, if any, and to what extent was actual science discussed?
- Watch the movie *Tree of Life*. What are the series of strong images and scenes that are fairly common at various stages in the film (*i.e.* the scenes that do not involve people)? What is their scientific background? What do they have to do with the story (*i.e.* what is the allegory)? After having watched the film, do you have a deeper appreciation of the scientific theory of natural history?

- Watch an episode of crime drama. *Sherlock, Inspector Morse/ Endeavor*, even *Doctor Who*, or similar investigative programs are good examples. Please state the program and the episode title. What are some scientific principles discussed and utilized to solve the case? Is the science necessary for the plot? Is the science accurate and correct?
- Watch the movie *Cloudy with a Chance of Meatballs* (the first one). How are scientists viewed at the beginning of the film, and how does that change at and after the resolution? How do characters' attitudes towards science affect their attitudes towards Flint? Why does Sam Sparks cover up her desires to be a scientist (meteorologist in this case)? Why does Flint ultimately prevail?

1.5.2 The Nightly News

Assignment for Students: Watch an episode of the national nightly news in which a science story is reported. It could be about engineering, medicine, pharmaceuticals, space, physics, biology, environmental science, *etc*. State what the story is about, and what day and on which news program it aired. Discuss in a two-page paper (the same as in week 1) what the science actually is and what made it newsworthy. Then, since you are a scientist, you will critique the discussion of the science.

2 Writing Science Through the Tenets of Journalism

2.1 Introduction

The most important part of any document is the first sentence. This is the only thing that the reader will read if he or she makes it past the title. This sentence must contain most, if not all, of the pertinent information in the subsequent news story; this sentence basically serves to summarize, synthesize, and advertise the entire document. This is often the opposite of actual literary story-telling or even standard scientific writing where the punchline is given at or towards the end of the piece. However, most human beings, scientists included, read in a journalistic fashion instead of a literary fashion without even knowing it. The title is interesting, the abstract gets a look-over, the introduction is started and quickly turns into being skimmed, and we wind up (30 seconds later) glossing through the conclusions for any nuggets to confirm or negate our initial judgements made from the title or abstract. We tell our graduate students NEVER to read a paper this way, but we do it all the time. Hence, we as scientists should not fight this natural human desire but should learn to employ this knowledge to our advantage. Scientists are often some of the world's most prolific writers, and adopting this model could make our writing more likely to be read.

In communicating science, the written word is the most permanent means of recording findings and musings on those findings. This writing is often done initially with personal notes that are only later refined into submissions for peer-reviewed publication. These

Complete Science Communication: A Guide to Connecting with Scientists, Journalists and the Public
By Ryan C. Fortenberry
© Ryan C. Fortenberry 2019
Published by the Royal Society of Chemistry, www.rsc.org

papers are often lost in the midst of thousands published every week. However, a re-evaluation of scientific writing style could alter this paradigm. While not sacrificing our need for correctness and clarity, the time has come for scientists to borrow the lessons of journalism and write for our scientific peers in a similar way that journalists write for their audiences. The relatively new American Chemical Society journal *ACS Central Science* even provides a template for abstract writing for its submission. This template follows the tenets of journalism perfectly. Former CBS Evening News Anchor Dan Rather has challenged scientists to adopt this type of communication model in order to provide audiences with more understandable information.

In order to develop this mindset of writing, journalism and even specifically science journalism, must be discussed first, and this is done in the present chapter. This discussion will serve both to instruct in popular science writing and then to be brought to bear in technical writing. In either case, the essence is to tell a compelling story but in a punchline-first format regardless of whether you think there is an audience for the piece or not. Sometimes, the story simply needs to be written, and it needs to be communicated in a means natural for information uptake, the style of journalism.

As a final note of introduction, the best writers are avid readers. They do not necessarily read things related to their medium, genre, or even type. However, they read. Hence, reading novels for fun will make a scientist a better writer. Reading political blogs for fun will make a sports journalist a better writer. Exposure to the written word as a consumer will be make for a better producer of the written word. The key is to enjoy reading, and that will make writing more enjoyable.

2.2 The Inverted Pyramid

The first sentence is the most important. This point should be clear by now. Then, the rest of the paragraph develops the background behind that sentence while the rest of the document further elaborates the story. Truly, the punchline is told, and then the rest of the joke is delivered. This model is called the inverted pyramid in journalism modeled in Figure 2.1. The foundation is at the top with the least important item at the bottom. Again, this seems counter to most of us are taught about setting, rising action, climax, falling action, and resolution (which is utilized in oral presentations discussed in a later chapter), but journalists have employed this model for more than a century to great affect.

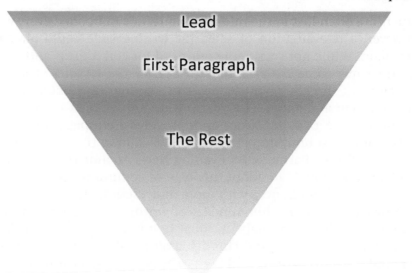

Figure 2.1   The structure of the inverted pyramid.

The lead sentence, or simply "lead," contains everything. Notably, it tells the reader why this story is important and why they should keep reading. Also, it should not tell everything. The lead should leave enough holes in the story so that the reader wants to continue to find out more information. A common misconception is that a good lead may ask a question. However, the writer should remember that if the reader answers "no" to the question, he or she will stop and not continue.

Compare the following two leads for the same story:

- "Have you ever wondered what molecules are floating around in space?"
- "The same molecules that make up our DNA are floating around in interstellar ice balls."

Most readers would answer the first sentence with a simple, "no." A few might stick around to read more because "space" is in there. The second is chock full of information. It relates the distant to the personal, describes the material in question, and raises a point that is begging to be developed. Even if the rest of the two articles were exactly the same, the one beginning with the latter lead is more likely to be read.

The rest of the first paragraph indulges the first sentence. In science communication/journalism, this is the point at which the researchers and where they work needs to be described. Furthermore, the journal article where the work is published also needs to be uniquely described (more of that later). Basically, a sentence after the lead should state something to the effect of, "A team of chemists at Northwestern South-Central State University led by Sir Hamilton Bright have published in the *Journal of Important Science* that . . ." Potentially, also some information about the experiment or observation should be provided, and what it all means, a type of secondary synthesis statement, should be placed at the end of this first paragraph in order to drive the rest of the document. Likely, at this point, less than 100 words have been used. Most journalistic pieces, especially for typical news, are 500 words. As a result, only 20% of the space has been taken up, but roughly 90% of the information should have been divulged. Now, what is done with the rest?

This is where the details start to be discussed. This can really contain any information necessary for developing what is known. Remember that even with a good lead, more than half of your audience has already stopped reading. The information that continues at this point should really be geared for anyone who is truly interested. Often these are stake-holders like watchdog groups, practitioners of the field, or those who make it their business to know these things. Hence, they really do want to know more about the topic you introduced in the lead and teased in the headline. Give it to them.

Each successive paragraph should be more and more disposable. Again, this is the guideline of the inverted pyramid. Furthermore, if a piece is too long, the editor will have to start chopping out information. The easiest place for him or her to do so is from the end where the fewest number of eyes will land. Do not put the most important information at the end. Do not put a summary statement. Again, that is the lead and goes at the very beginning. In journalism, the piece really just fizzles out. There is no flourish. Journalism is the opposite of Neil Young's opinion, it truly is better to fade away.

The second and following paragraphs can contain thoughts from the scientists who performed the research as well as thoughts from scientists familiar with the research but were not involved in the actual work. More detailed descriptions of the work might be included here. How the test subjects were chosen or what were some interesting hiccups or side insights are part of this development. Everything after the second paragraph is to fully round out a story, but it should, again, go in successive chunks of increasing uselessness.

2.3   The 5 Ws of Journalism

The lead, the first paragraph, and the rest of the article within the inverted pyramid model have to answer the standard five Ws (questions) of journalism: "Who?," "What?," "When?," "Where?," and "How?" While the last does not start with a "W," the nemonic is made better by the outlier. Plus, "how" has a "W" at the end. The lead should answer as many of these questions as possible.

Returning to the example lead above: "The same molecules that make up our DNA are floating around in interstellar ice balls." In this case, "Who?" is not as important as "What?" and will be left for the rest of the first paragraph since "Who?" are the actual readers and, while not as important as "What?," they are inserted into the lead for immediate common ground. "What?" are the same molecules that are in our bodies. "When?" is implied as pretty much always. "Where?" is in space, which is most often given a notable "cool" factor by most readers. "How?" is left out, purposefully. People naturally ask these questions and want them answered. Hence, they should continue reading since a full, but unsatisfying, summary has been provided in the lead.

Making a strong case for the rest but leaving out the "How?" is one means of providing all of the necessary information but not providing ALL of the information as a good lead should. Aristotle's logical syllogism in the form of: "If A then B and if B then C, then if A then C," is easily modified to remove one of the two premises leaving a hole—called an enthymeme—for the audience to fill. This hole left to be filled in by the audience results in self-convincing. Consequently, setting up a complete piece of information but keeping back a bit that makes the reader want to continue will get him or her to read the rest of the first paragraph. This must be done with care, as it can also confuse the reader. When in doubt, however, give away all 5 Ws.

The rest of the first paragraph answers the Ws not given in the lead and it also develops those given in the lead a bit more. "Who?" in the case of our DNA snowballs in space is the audience and is nominally also about who did the research. However, it can also be about who is affected by the research or by whom the research was funded. Any people, person, or public aspects are included here.

"What?" is what was done or, also, what it means. Putting "What?" in front of any objective clause imaginable helps to develop the story fuller as "What?" is so open-ended.

While the "When?" in this example is implied to be since the dawn of creation, the discussion that should come at some point in the rest of the first paragraph about how a million years worth of photons were pounded onto an ice ball develops the "When?" to a further degree. Additionally, "When?" this research was done is an important aspect of answering that question. If it was done 12 years ago, it likely is not important any more. The concept of newsworthiness is discussed below.

"Where?," of course, could be where the research was done or where the results have the most impact. This example, the "Where?" is in interstellar space AND the reader's own body. If the story were about unsafe drinking water, the specific example could be the locality where this was discovered but also with the implication that the "Where?" could be in the reader's own kitchen. Making the "Where?" both objective and somewhat personal to the reader makes them more interested unless they are already practitioners in the field.

"How?" is, naturally, the crux of the science. As a result, in science journalism, most of the focus will be on "What?" and "How?" Even so, the science journalist cannot neglect making it personal in some way. Really, the personal aspect of the "How?" and "What?" lead to another "W" largely for science journalism. "Who cares?" This must be answered first, or the article has no business being written. Often, "Who cares?" is taken care of by the public relations office of the journal, research institution, or editor, but knowing whether something is or is not important as the scientist doing the research or the journalist writing the story can be as significant as the "When?" being timely or too long ago.

The combination of the "When?" and "Who Cares?" is the idea of newsworthiness. Great discoveries in science are often only great to those working in the field. This is one of the major disconnects of science and journalism and why the two groups often miscommunicate with one another. The journalist is always looking for *THE* story. It can often be quite difficult to separate the incremental from the truly important which is why new cures for cancer are reported regularly. While reporting of these incremental steps is of vital importance in a field as personally involving as cancer research, it truly must be a cure before the story can really be written and published. The audience of the medium will also often determine whether something is newsworthy. Most often, newsworthy stories for popular and broad-based consumption (not specialized audiences) will be directly impactful for personal welfare or are fun, fascinating bits of human insight. Finding the cause of autism is the former, while discovering flaming ice balls is the latter.

2.4   Style Including the 3 Cs of Science Journalism

Journalistic writing is written in the third-person, present tense, active voice, omniscient viewpoint. Additionally, there are "3 Cs" of science journalism. Any piece must be clear, concise, and correct. While personal style is absolutely essential in any writing, especially science journalism, these two precepts of style must be obeyed, or the text can become confusing, irrelevant, or biased.

Any personal pronouns besides "he," "she," "it," or "they" should be avoided. Rarely personal accounts are acceptable, but these will not be the standard form. The journalist is the impassive observer serving as a fiduciary for the reader who could not possibly observe all of the events reported. The reader must take the information given to him or her from the journalist who wrote the piece and then make up his or her own mind. The journalist provides the information. The reader provides the interpretation. The journalist must not provide interpretation. While some bias inevitably creeps in, this should be minimized as much as possible. The observations should be reported as true to form as humanly possible.

Journalism, again, serves as the eyes and ears of the reader who cannot possibly be present for the various things being reported. Hence, the text must be in the present tense and in active voice. Present tense *IS* important as it makes the reader feel as though he or she is experiencing the action as it happens. Sometimes, past tense can be used when describing an event that has already passed, but this should be done only when sensible and not as the default. Active voice seems confusing but is really simple: Put the subject at the beginning of the sentence. An easy rule of thumb is that no sentence should begin with "It is." The antecedent to "it" has not been defined yet and can only be in the predicate. "It is cold today" is a form of passive language. Instead, "Today is cold" is much more active. If the thing in the sentence doing the stuff is first announced after the verb, passive voice is being utilized. The object performing the action must always precede the verb within the sentence structure.

The omniscient viewpoint is simply that the reader feels as though the journalist author of the piece knows all and sees all. Gaping holes in the story should not be present. All of the necessary information must be contained within the piece such that the reader does not feel compelled to look elsewhere for information. Any such information should be properly credited. Finally, the document should give a complete description of the item or action being reported.

Beyond these things, science communication has its "3 Cs" of clarity, correctness, and conciseness. Clarity stems in large part from the above tenets of journalism itself. Deeper development for science journalism has other factors, most notably including the avoidance of jargon. As mentioned in the previous chapter, scientists have evolved their own language to describe their subject matter in as few symbols (words) as possible in order to avoid confusion and reduce communication time. Some of these concepts have to be unpacked by the journalist. For instance, vocabulary may have to be defined or described with "reducing agent" being replaced with "molecule in the reaction that provides electrons." Often analogies are useful in describing difficult concepts and can be fun. "Fluorine is highly electronegative" could be easily described with the analogy, "Fluorine is like that kid on the playground who never lets anyone play with his or her ball. Fluorine steals the other kids' electrons and will not easily let them go." These can be fun, but they can be too distracting. Balancing between edutaining and distracting can be difficult but is mastered through practice and consulting with an editor or someone who can read over work before submission.

One note on the use of jargon, especially within the chemical sciences, is that some words utilized within chemistry are in many ways the opposite of what common sense would indicate them to be. Furthermore, there can be many names for the same thing. One example is a diradical, triplet state, or double spin-unpaired system. In fact, science writer Lee Billings in his book *Five Billion Years of Solitude* even calls out chemists for their obtuse usage of jargon saying, "Confusing vagaries of nomenclature ... are in large part why many science journalists avoid writing about chemistry." Consequently, the chemical science communicator has to find novel ways of translating the years of tradition within the language of chemistry into the common vernacular.

Being concise in science journalism is not always easy since many concepts are difficult. However, walls of text or even the best analogies, if drawn on too long, will lose the audience. The ability to clearly describe information in as few words as possible makes them more likely to be read, first of all, and, second, more likely to be understood. This also extends to sentences. Shorter sentences are easier to process. Periods are more useful than commas. If the reader feels as though he or she can chop up what is being read into smaller bites, he or she will more easily make it through the text.

Correctness should be self-explanatory. The last thing a science journalist should be is incorrect in what he or she is reporting. Without correctness, there is no point in doing this. Reporting falsehoods are worse, frankly, than not reporting truths. Improper reporting can lead to confusion at best or anti-science at worse. All journalists and writers make errors. Again, having someone else critically read a piece, especially a human source (discussed next) can also aid in strengthening correctness.

2.5 Sources

All reported information comes from someone besides the journalist. This is the source. The scientist engaging in the research, a scientist engaging in the field but not necessarily involved in the research who can add commentary, and the peer-reviewed journal article in which the research appears are really the only primary sources. Everything else is pretty much a secondary source. Only primary sources are acceptable for journalistic writing. At least two sources are needed for standard journalistic writing of 500 word stories. Longer pieces require more sources, typically.

The primary source is intimately involved in the research. She, he, or it has the necessary information for the journalist to distill to the reader. The easiest place to find a story is either in the scientific literature or from a scientist with whom the journalist has a personal relationship. In the case of the former, once a research article is found, the author of the article (since there is always a corresponding author) is a second, primary source. For the other case, the known scientist can point the journalist to another scientist or one or more journal articles that can be the second primary source.

The scientists themselves are often the best sources. They can explain and describe what is happening. They can answer questions or offer corrections. They also give a human "face," so to speak, to the research being done that is often described with bland numbers or raw figures. The connection between the reader and the scientist source can also make the piece of human interest and give the reader a sense of relationship with the research practitioner. Quotes are necessary when writing from people sources, but any more than two from a single source (unless that person is the subject of the piece like in a personal feature article) can be gratuitous. The journalist's job is to distill a majority of the quotes into digestible chunks for the reader and not simply quote the more removed scientist. The first time a name, especially of a people source, is given should include the first name, last name, and appropriate title; thereafter, simply the last name is sufficient.

The peer-reviewed article is often the best place to hunt for a story. Top-tier journals like *Science* or *Nature* have articles that are often in the newsworthy category simply by the nature of their publication practices. However, more discipline-specific journals can be as well, such as *Angewandte Chemie, Chemical Science* or *Chemical Communications*. The articles can provide the necessary raw data to be discussed and digested by the journalist for the reader. Additionally, they can provide other human sources in the form of the scientist authors of the citing or cited works from the article of interest. Most often, the peer-reviewed article is the final word in any scientific advancement since the data are now available in perpetuity. The work has been peer-reviewed and approved by at least one reviewer and one editor.

Additionally, building a relationship with the scientific or managing editors of journals can also provide a primary source. Conflicts of interest can arise, however, since promoting of a story within his or her journal makes the editor look good. Even so, the editor can potentially point the journalist to stories of note that may not be easily seen or to other scientists who can be the primary source.

Secondary sources include other news articles, press releases, encyclopedias, and anything else that did not directly "observe" the experiment in question. While these can lead to ideas for good stories, none of these are acceptable as primary sources. Secondary sources may contain errors or lead to misinterpretations in the material being communicated. Such violates the "correct" of the "3 Cs" of science journalism. When in doubt as to whether or not a source is primary or secondary, if the source is not a person or the peer-reviewed article, it should be treated as a secondary source and not used in the story.

2.6 Interviews, Interviewing, and Being Interviewed

Since people are a primary source, the journalist must talk to them. When interviewing a scientist, the journalist should, first, do all of the things necessary when doing any interview: (1) have a list of prepared questions; (2) do background reading on the story in question, related material, and the communication surrounding the material and interviewee; (3) look-up background on the interviewee (previous employment and education, most notably); and (4) meet the interviewee at a time and place most convenient for the interviewee.

Additionally, the science journalist has to be able to converse in the language of the scientific area in question or at least be competent to ask informed questions. Finally, the journalist needs to acknowledge the differences between himself or herself and the scientist. Scientists are people and have personalities across the spectrum. The introverted stereotype can be true, but scientists are unique individuals and should be treated as such. Mostly, the interviewing journalist needs to be interested in what the interviewed scientist wants to say, whether it is on topic or not, and should allow the conversation to progress naturally utilizing the prepared questions as an outline and not a strict road map.

Since this text is being pointed towards scientists, the interviewed scientist also should remember some items about the journalist, should the opportunity to be interviewed arise. First, journalists are people, too, and have jobs to do. They are typically not out to sensationalize things. As a result, the scientist needs to clearly communicate the science story being developed by the journalist and respond to the journalist's questions with as much grace and clarity as possible. Often, misunderstandings in journalism arise when the two parties, interviewer and interviewee, talk past each other. Even when being interviewed, the scientist should listen to the journalist. The scientist should also be mindful of the journalist's time since he or she is likely working on multiple stories at once just as the scientist is likely working on multiple research projects at once.

In all, both parties should view the other as human beings and then as experts in their chosen fields. This level of collegiality is necessary for productive conversations since interviews really are conversations with a purpose. Finally, the journalist should provide the scientist with a chance to review the piece before publication in order to correct any misconceptions, which, again, helps to strengthen the correctness of the 3Cs. The scientist should do so in a timely manner, and the journalist should provide the piece to the scientist with some lead time. Sending it to the scientist at 5 o'clock on a Thursday with publication at 8 the next morning is probably adequate from a journalist's perspective, but it may not be for the scientist.

2.7   Length, Audience, and Media: Old and New

Most journalistic pieces or stories are right at 500 words. This often equates to a five-paragraph essay. In holding with the inverted pyramid, the most important words are the first 12 to 20 in the first sentence followed by the next 80 or so to complete the first paragraph.

Hence, most readers will only ready the first 100 words or less of a story. While science journalism and journalism in general can provide up to 500 words, really, the story must be told in the first 100.

Feature pieces can be longer, especially in magazines. These can top out at 10 000 words in some cases with most occupying the 2000 word range. Longer pieces often require more elements of traditional storytelling and more sources, but, again, most readers will only make it through the first paragraph.

The length of any piece is often determined by the medium which is then often determined by the audience. Older generations still prefer printed materials. Younger generations are electronic only. For now, most consumers will land somewhere in the middle, until the younger generations begin to dominate the media consumer markets. Knowing who reads the medium for whom the journalist is writing will determine the way in which the information is communicated. More technical language should be utilized for a trade-specific outlet even if it is journalistic in nature. As such, longer stories may be natural for such media. The converse is also true in general news outlets. More concise stories are appreciated by readers when they are non-experts. Being able to get the nugget of information and get out is key.

The most easily conceived concise story is a tweet. However, journalists have been writing tweets for centuries. Tweets can simply be viewed as leads. Hence, when writing tweets for NASA, the tweeting journalist will have a mixed audience in terms of expertise, but most will be somewhat conversant in astronomy and space flight. Tweeting for the BBC is the general audience extreme, while tweeting for the International Society of Electrochemistry would be the other. The audience determines the type of tweet even it if is the same information.

Additionally, electronic or "new" media are most often multi-platform/multimedia. Now, even when news stories are written, they are posted online with videos, graphics, and hyperlinks to related stories that augment the piece. Furthermore, social media promote the story, often through tweets or pins. Hence, the modern science journalist must know whom his or her target audience is, how they will most likely access or find the story. Then, they must also consider (often with the help of an editor or a specialized expert) what additional, multi-platform electronic objects can support the piece. As a result, the science journalist must own and promote the entire experience for the story being developed from all angles. As electronic media dominate, this will only grow in significance.

2.8   Assignments

2.8.1   The Lead

Writing the lead is often the hardest part. They are usually too long, too vague, or too grandiose. Clear, concise, and correct are key.

Assignment for Students: Write a lead for the research project that you are doing now.

Then, remove two words.

Now, remove two more words. This may require you to rewrite portions of it.

Now, remove three more words.

Share with the class who has the shortest lead. Then, judge together whose is still the most coherent and the most catchy. Rank those. Add those ranking numbers with the number of words in the lead. The person with the lowest score wins. Give them a prize.

2.8.2   First Journalistic Style Piece

This is meant to be a low-barrier assignment just to get students in the mindset of writing journalistically. This will be a bit of a leap for many of them. Be advised that most will still try to write a concluding sentence, the lead will likely be vague, and sources will be utilized improperly. Also, encourage them to read the science sections of online news sources every day. Finally, extra credit for using a people source is a great way to motivate students to talk to experts who are often the best resources.

Assignment for Students: Find a recent issue of *Chemical Science*, *Nature*, or similar wide-range chemistry or general science journal. Select an article from there that interests you. Remember, you are not getting a news article or an opinion piece or a review. You are getting a piece of peer-reviewed work. You are using a new, fresh report of interesting scientific findings. This should be clearly listed in the table of contents. Ask me if you are not sure if your basis article is proper for a peer-reviewed article. Using what you have learned from our discussions, write a 500-word news article. The document should be double-spaced. At the top of the page, simply put your name in the upper-left hand corner. A line below that, put what you think would be a good headline for your article. It should NOT be the same as the journal article's. Then, on the line below that, begin your story. Use an inverted pyramid, tell a story describing the research and why it's important, and follow as many of the tips from the texts as possible. Attach the journal article you selected to your story or put a reference (in RSC style) on a separate page.

You must use two sources. An easy place to get a source is a reference in the paper or a paper that cites the one that you chose (if one has already come out). You must cite both sources in your story in AP style. This is different from RSC citations. You would simply write something like: "In the November 29 issue of the journal *Nature*, Fortenberry and Magers discuss. . . " for a journal article or "According to scientist Jack Jackson at IBM's Almaden Research Labs. . . " for an interview or conversation.

You will receive 50% credit for simply turning in the assignment. Provided that the story is within 500 words (but not too many short of that), about scientific research you found, and the journal article is either attached or there is a reference, you will receive the other 50% credit.

2.8.3 Peer-review

Post the previous assignment anonymously to an online resource. The other students must read their peers' papers and offer praise and criticism for what was written. This can be done in class where public discourse can lead to new ideas. Alternatively, it can be done as a written assignment with directions below. In this case, post the reviews (also anonymously) after they are due. The key is doing these in close proximity to one another. Posting a peer-review any more than two weeks after the initial assignment deadline is not beneficial.

Assignment for Students: Select and read four of your fellow students' anonymous articles not including your own. Write a ~400 word criticism of each. At the beginning of each section, write the title of the piece. State what you thought your fellow students did well and what they did not. Be constructive but polite. Be honest but not cruel. Since each of you are learning this process together, peer feedback is the most important since you each know where the other is much better than the instructor. Remember, others will be doing this for your work. You will be benefit from their constructive comments so make constructive comments for them.

2.8.4 Second Journalistic Style Piece

Assignment for Students: In this article, you are to write a 500 word (maximum) story discussing a recent news item of scientific significance. Remember to use two PRIMARY sources. Both can be journal articles or discussions with actual scientists, but you cannot quote a magazine, newspaper, or other news outlet as that is a secondary source. Build on the comments and criticisms of your peers and the instructor from last week's *Chemical Science* journal-inspired article in your writing. You should use the tricks discussed to come up with an idea about a science story that is relevant to the current time, and also to write your story. You may want to discuss the FDA's approval/rejection of a new drug, a recent result from the Large Hadron Collider at CERN, the newest alternative fuel research, how the latest economic forecast was developed, *etc.* Remember that this is a news story you would want to submit to a newspaper for everyone and your mother to read. Like last week's assignment, again double-space the document, put your name in the upper-left hand corner with the headline underneath, and then begin the story.

Grading of the assignment will be based on the following criteria:

- Simply turn in the assignment (50%)
- Generate a story idea that is about science (20%)
- The quality of the writing both through the use of proper English and clear science journalistic communication (30%)

2.8.5   Interview Article

Assignment for Students: Each student is to interview a scientist for a 1000 word feature story which is geared towards appearing in the University magazine. This story must be about the scientist's research, but personal factors can be (and often must be) included. Go to the interview (or the call if it is a phone interview) prepared with questions. You should have some idea of what type of story you want to write before you start the interview. That will seed your interview. Get to know the scientist's research beforehand so that you can ask intelligent, informed questions and follow along with the discussion. However, you may get exposed to new things, and feel free to work "off-script" if necessary to get a good or better story out of your interview. Again, you must have two sources. The second could be from a fellow scientist, competitor, a student in his or her lab, a journal article citing his or her work, *etc.* In this exercise you are getting information straight from the scientist. Hence, you will be communicating scientific concepts to the "masses" straight from the horse's mouth (so to speak). You may not interview a scientific faculty member whom you have had as a research mentor or director. It must be someone who does not know you on a deep personal level.

Grading of the assignment will be based on the following criteria:

- Choice of scientific topic (10%)
- The use of two sources (10%)
- The scientific qualifications of primary interviewee (10%)
- The use of explanatory and narrative skills (10%)
- Clear, concise, and correct writing (60%)

2.8.6 Story Idea Meeting

Most newsrooms have a meeting every morning (or weekly depending upon the periodicity of the publication) in order to discuss what will be included in the next news cycle. This assignment is to simulate that environment in order to have the students contribute to and participate in such an exercise. This would a good "Participation" grade. Otherwise, the criterion of have two story ideas can count as 50% credit and making useful contributions to the discussion would be another 50% credit.

Assignment for the Students: Talk to people within your department. Get an idea for items that are newsworthy. What is going on? Who has had a big research breakthrough? Who has won an award? Is there are a departmental club that is doing an event? Find out what "new" information is floating around in order to make a news story that will flow with the "Final Article" described next.

The instructor will function as the "Editor" of the department news website. Bring two story ideas to the class meeting where the "Story Idea Meeting" will take place. Some of you may have the same ideas which is why more than one is necessary. There are no bad ideas. The "Editor" will write all of the stories on a board or projected screen. Then, you will discuss as a group which are the most important stories. The "Editor" will rank these with your inputs. Then, suggestions will be taken for the most important stories. If there are no suggestions, the "Editor" will assign the story at his or her discretion.

2.8.7 Final Article

This assignment represents the culmination of the "Journalism" section.

Assignment for Students (Major): You are to develop a story idea that will go on the news portion of your department's news website and write the 500 word piece. You need to think of the proper audience, news story, and newsworthiness for this venue. You will need to include at least one photograph. Your instructor can help you vet ideas, but he or she will not help you to come up with them. That is part of the assignment. You must have two sources, and one must be a real, qualified person with whom you have spoken who is not the same as one you interviewed in the previous week's assignment.

Grade assignment will be based on the following criteria:

- Choice of topic (10%)
- Use of sources (20%)
- Explanatory and narrative style (10%)
- Clear, concise, correct writing (60%)

2.9   Examples

2.9.1   The Intersection of Biology, Chemistry, and History

Tree Rings Tell a Tale of Wartime Privations
By Bas den Hond
Eos, *99*, April 11, 2018

"Where is *Tirpitz*?"

Britain's prime minister Winston Churchill, who was hell-bent on seeing Germany's largest battle cruiser destroyed, asked the question in 1942 in a memorandum that was notable for its shortness. More than 75 years later, German forest ecologist Claudia Hartl is able to give an answer notable for its method: the growth rings of pines on the Kåfjord in the far north of Norway show that the *Tirpitz* was anchored there in 1944.

Hartl of the Johannes Gutenberg University in Mainz, Germany, was not looking for an answer to the *Tirpitz* question when she stumbled upon it. She was investigating instead why something was amiss with some trees she and her students happened to be studying on the Kåfjord. In contrast to the normal-seeming tree rings from other locations in northern Norway, a few cores the researchers took in this area showed no growth ring or a ring that was hardly visible for the year 1945.

Today, at the annual General Assembly of the European Geosciences Union in Vienna, Austria, Hartl discussed new findings that German attempts to hide the massive warship from enemy attacks by means of a chemical fog left enduring evidence in the trees. The previously undiscovered evidence consists of varying amounts of damage to long-lived Scots pine trees on the lands around Kåfjord. This forensic research is, as far as Hartl knows, the first example of "war dendrochronology." The team has plans to use additional methods to trace that wartime history in the affected trees.

Hartl and her class discovered that something was wrong with the trees at the Kåfjord after a 2016 excursion to Norway. In subsequent years, they followed up and learned that trees nearer to the fjord had skipped even more years. Some nearest the fjord had even stopped growing for as long as 7 years, returning to normal only after 12 years.

Often, such patterns are explained by drought or insect attacks, but the trees that Hartl had analyzed, Scots pines, are too hardy to stop growing completely in those circumstances. "It is really unusual for Scots pine in Norway to miss a ring," said Scott St. George, a climate dendrochronologist from the University of Minnesota in Minneapolis who spent a year in Mainz as a Humboldt Fellow and participated in the study. The absent ring indicated that the tree was going all out to survive. "Because it was not forming wood around its circumference, it was taking all of its resources and all of its energy to regrowing a complete crop of needles from top to bottom," St. George explained.

"Insects don't affect them to that severity; a cold summer does cause them to form a narrow ring, but to have 60% of all trees not form a ring at all, that's really odd," he said. Furthermore, according to Hartl, insect outbreaks that damage trees "have cycles, so you will see missing rings every 7 or 10 years. This is well known for the Alps but not that you have just one single ring, or several, missing over 200 years. That's really uncommon."

She and her colleagues wondered if there was something special about the Kåfjord. One day, she asked a Norwegian tree ring specialist, who replied immediately that the *Tirpitz* was there.

The largest battle cruiser ever built for the German Navy, the *Tirpitz* saw little action, but it was essential in forcing the British navy to deploy resources to prevent it from attacking convoys. While it was stationed in Norway, it was repeatedly attacked by submarines and especially by bomber aircraft. To make the *Tirpitz* harder to hit, the Germans released chlorosulfuric acid from the ship and from land. Droplets of this compound attract water, forming an impenetrable mist in a matter of minutes. British bomber pilots said that the fog covered the terrain to a height of 600 meters.

Hartl and her research team got their information about the cloaking fog from American military reports. Chlorosulfuric acid is irritating for people, but it was otherwise described as harmless at the time, St. George said, "because cows exposed to it didn't die immediately." But for pine trees the effect was severe: Their tree rings show that they lost most of their needles.

When St. George and Hartl speak about the trees at the Kåfjord, it is with a certain admiration.

"A tree which shuts down over 9 years, but it's still alive, imagine this," Hartl told *Eos*. "That tree is amazing."

Regarding this new study, "I found it pretty cool," said Georg von Arx, a tree ring expert at the Swiss Federal Institute for Forest, Snow and Landscape Research in Birmensdorf. "It was very original research, and it shows what a huge diversity of questions we can tackle with tree rings. And scientifically, it was very well done."

Von Arx agrees that the missing rings are a sign that the trees had lost their needles. And he offers advice: "I think if she has the chance she should go back and sample them a bit higher up. I would expect to see fewer missing rings as you move upwards in the stem." This is because a growth ring consists of many rings of sapwood, which transports water. "In the sapwood at breast height you still have alternative pathways for the water, but as you move towards the tips of the branches, there are fewer and fewer sapwood rings, eventually only one. And a new needle needs to be connected to a new level of cells: either there is a sapwood ring there, or there can be no needle growth. This difference with height would show that the tree was rebuilding the canopy before—and this is always second priority—growing the stem."

The *Tirpitz* was eventually sunk in November 1944, but its legacy remains in the trees, which Hartl said can survive 400–500 years. It is possible that the acid fog has left not only physical traces in the tree but also chemical ones inside the wood. The team is now looking for evidence of that, analyzing their samples with a mass spectrometer.

Finding out as much as possible is important, according to Hartl, because until now nothing was known about the environmental consequences of battles involving the *Tirpitz*, and the same goes for many other engagements in wartime. "The Kåfjord was not the only place where the German navy used the smoke. They used it to obscure other ports and other cities. There may be fingerprints in other places that people haven't paid attention to," said St. George.

In some sense, the Scots pines of Norway are also monuments, he added. "We're losing the generation that remembers the Second World War. It's easier to forget the lessons after that event. In this case, the trees still keep that evidence alive. Even if you didn't know anything about the history of the Kåfjord, or the *Tirpitz*, the information is still preserved, when you have the right person to read it."

2.9.2 Honey Bees and Cerium Oxide

Summer Bees Vulnerable to Nanoparticles in Fuel Additives
By Hannah Kerr
Chemistry World, October 26, 2017

Cerium oxide nanoparticles cause biochemical changes in honeybees. Scientists in Slovenia have found that two enzymes associated with learning and toxin removal are disrupted in bees that consume a common catalytic nanoparticle.

"Honeybee population decline is a worldwide important issue," says Anita Jemec Kokalj from the University of Ljubljana, who worked on the study. Recently, pesticides have taken the blame for this decline and EU regulations on neonicotinoid-based insecticides look set to tighten. Cerium oxide nanoparticles, however, may also be harming bees and scientists have already found that they induce biochemical changes in other insects. Cerium oxide nanoparticles are "widely used as a fuel catalyst and it also has many other applications. Therefore, it can be deposited on plants and come into contact with pollinators," explains Jemec Kokalj.

The biochemistry of honeybees born in summer is quite different from those born in autumn. Worker honeybees tend to live for only 40 days in the summer months, but autumn bees usually stay in the hive and live through until the next spring because they do not raise any young over the winter. This means that winter bees have slower metabolisms, while summer bees have different levels of enzyme activity to promote learning, which is important for foraging.

The Slovenian team fed worker honeybees collected in both summer and winter with sucrose solutions spiked with cerium oxide nanoparticles and measured the activity of two enzymes after nine days of chronic exposure. The bees did not die from eating cerium oxide nanoparticles, but summer bees that received a $250\,\mathrm{mgl}^{-1}$ dose of the nanoparticles became visibly hyperactive and agitated. The team observed a general increase in activity of one of the enzymes, acetylcholinesterase, which is involved in honeybee learning. This indicated that the central nervous system of the bees was affected. They also observed an increase in the activity of an enzyme associated with pollutant removal. Bees collected in the summer months were significantly more affected than those collected over autumn.

"This is an important study," says Jason White, vice director of the department of analytical chemistry at the Connecticut Agricultural Experiment Station, US, who was not involved in the work but has investigated CeO_2 toxicity in plants. "The different response of the summer and winter bees is notable and suggests potential species-level differences as well."

However, accurately measuring ecological concentrations of cerium oxide nanoparticles is challenging. Ryszard Maleszka, an expert in how diet affects bee behaviour from the Australian National University, points out that in general "chronic lab testing… is not necessarily representative of environmental exposure."

"Non-lethal yet toxic responses at [low] dose are of significant concern … additional mechanistic studies that include non-nanoparticle controls should be undertaken," adds White.

Most of the nanoparticles remained in the bees' digestive system so Jemec Kokalj was surprised to see any effects "because the nanoparticles were most probably not translocated from the gut to other body regions. Our next step is to investigate whether the nanoparticles do translocate to other body regions."

2.9.3 Malaria Work Restarted

Global Leaders Seek to Reignite Fight Against Deadly Malaria
Voice of America/VOA.com, April 17, 2018

Renewed action and boosted funding to fight malaria could prevent 350 million cases of the disease in the next five years and save 650 000 lives across Commonwealth countries, health experts said Wednesday.

Seeking to reignite efforts to wipe out the deadly mosquito-borne disease, philanthropists, business leaders and ministers from donor and malaria-affected countries pledged 2.7 billion pounds ($3.8 billion) to drive research and innovation and improve access to malaria prevention and treatments.

Spearheaded by Microsoft co-founder and philanthropist Bill Gates, the leaders warned against complacency in fighting malaria—a disease that kills around half a million people, mainly babies and young children, each year.

While enormous progress has been made over the past 20 years in reducing malaria cases and deaths, in 2016, for the first time in a decade, the number of malaria cases was on the rise and in some areas there was a resurgence, according to the World Health Organization.

The disease's stubbornness is partly due to the fact that the mosquito that transmits the disease and the parasite that causes it have developed resistance to the sprays and drugs used to fight them, health experts say. It is also partly due to stagnant global funding for fighting malaria since 2010. Climate change and conflict can also exacerbate malaria outbreaks.

"History has shown that with malaria there is no standing still–we move forward or risk resurgence," Gates said in a statement ahead of a "Malaria Summit" in London on Wednesday.

His multibillion-dollar philanthropic fund, the Bill and Melinda Gates Foundation, which is co-convening the summit, pledged an extra $1 billion through to 2023 to fund malaria research and development to try to end malaria for good.

"It's a disease that is preventable, treatable and ultimately beatable, but progress against malaria is not inevitable," Gates said. "We hope today marks a turning point."

The malaria summit was designed to coincide with a Commonwealth Heads of Government Meeting (CHOGM) in London this week. The 53 Commonwealth countries, mostly former British colonies, are disproportionately affected by malaria—accounting for more than half of all global cases and deaths, although they are home to just a third of the world's population.

Among new funding and research commitments announced at the summit, the Global Fund to Fight AIDS, Tuberculosis and Malaria said $2 billion would be invested in 46 countries affected by malaria between 2018 and 2020.

Pharmaceutical firms GSK and Novartis also increased investment into malaria research and development—of 175 million pounds ($250 million) and $100 million, respectively. And five agrichemical companies launched a joint initiative to speed up development of new ways to control mosquitoes.

2.10 Student Examples

The following examples were written for release on the Georgia Southern University Chemistry Department website within the "News" subsection as part of the Major Assignment for this section of the course. Note the scope and application of the pieces and how they are targeted for their audiences.

2.10.1 Award Notification

Professor Wins Prestigious Rising Star Award
By Ian Byrd
February 24, 2016

A Georgia Southern Chemistry (GSU) faculty member, Dr. Karelle Aiken, has received the Women in Chemistry Committee Rising Star Award for 2016. This American Chemical Society award recognizes outstanding early- to mid-career female chemists across the nation.

Aiken was one of only ten women hand-picked for the award due to her excellence in research and her passion for mentorship of all levels of students and even other faculty.

Aiken's drive to inspire and mentor others sets her apart and makes her an ideal candidate for such an award. This drive started early in her academic career. Ironically, an experience with a drama teacher at the high school she attended in her native country of Jamaica is what helped her realize her passion for helping others to reach their full potential. "She brought out things in me I didn't know I had in myself," says Aiken. Aiken uses that experience to guide her mentorship to others and help them get the most out of their experiences. When speaking with her current graduate student, Richard Govan, the pivotal role that she has played in his career is obvious. Govan noticed her passion while taking her Organic Chemistry course, which inspired him to change his career path and pursue chemistry.

Aside from her mentorship and teaching, Aiken is also responsible for leading the summer undergraduates research experience program at GSU, which is funded by the National Science Foundation. She is also a founder of the GSU chapter of the National Organization for the Professional Advancement of Black Chemists and Chemical Engineers. To add to her list of responsibilities, she has as many as five undergraduates working in her lab as well as one graduate student. All of these extracurricular activities played a pivotal role in her reception of the Rising Star Award. Her obligations may seem endless, but a burning passion for her job gives her the energy to keep moving.

Being truly passionate about a job is something that many seek but not everyone finds. Her initial plan was to attend medical school when she was living in Jamaica. After seizing upon the opportunity to start her education over at Williams College in Mass., Aiken had a change of heart. While on a physics track, she took an advanced synthetic chemistry class that led her to realize her passion for organic synthesis. This experience is the basis of her current research projects, many of which involve method development.

While being the recipient of such a prestigious award is very humbling to her, she does not plan to let herself get complacent. Aiken says she plans to keep evolving her research to keep herself on the cutting edge as well as keep forming strong relationships with students so that one day they will be able to pass on their knowledge to others, just as her mentors did for her.

2.10.2 Chemical Spill

Chemical Spill in Lab Leads to Minor Burns
By Oladayo Ariyo
February 16, 2018

Four students were involved in an acid spill in the chemistry department last week. The incident occurred in a CHEM 1152 Organic lab on February 7th. The lab was a separation experiment which involves strong and weak acids. The freshmen students involved were waiting for a sample of hydrochloric acid to isolate when one of the girls walked into another girl who was holding a beaker containing the acid. She dropped the beaker which caused its contents to splash on all three girls around the area. The fourth girl picked up a bottle with her bare hands not knowing it had some acid dripping on the side of the bottle. She was injured, as well.

Kayla Anderson, a first year graduate student, was the graduate teaching assistant (GTA) in the lab at the time of the accident. She assisted the students to the sinks to wash the acid off their skins and wrote an incident report for the department. Dr. Nathan Takas, the chemistry instrument manager was nearby when the incident occurred and helped the students to the health center for more treatment. Similar incidents are a frequent occurrence in lab courses where students have to handle corrosive chemicals.

Aaron Quarterman is a second year graduate student, and has been a TA in many organic labs. He has seen his share of laboratory accidents. When asked about the incident he felt that students are sometimes inexperienced when it comes to handling chemicals in laboratory settings. Quarterman also felt that protective attire such as lab coats were not adequately enforced by the department. When all students own a lab coat and wear it to labs, especially those where corrosive chemicals will be used, it ensures that these kind of accidents are reduced. Lab coats are enforced in other institutions as it encourages safety and proper laboratory practices.

Dr. Amanda Stewart, an assistant professor and faculty director for lab safety had some interesting thoughts on the accident, "Minor accidents involving broken glass, hot glass, and minor burns happen once or twice a month." Stewart said that labs are currently designed to prevent major accidents by minimizing chemical toxicity and concentration. She feels lab coats add a level of safety hazard because students do not always get them in the right size and large sleeves might be clumsy to deal with.

Stewart offered some advice for students working with chemicals for the first time. To prevent spills in the future, students should ask for gloves when handling chemicals and should always mix chemicals in the hood. Students should also avoid congregating in pathways, so as to prevent any clumsy collisions and follow instructions given by the GTA and instructor. GTAs and instructors should ensure students are aware of what might happen before starting the lab, prelabs should be focused on pointing out possible accidents and hazards.

In an unrelated incident in the same lab, a hot plate caught on fire when somebody left it on high heat. Thanks to the quick thinking of the GTA, the fire was put out.

3 Writing Technical Science Like a Journalist

3.1 Introduction

The first sentence even of a technical piece should do more than get attention; it must inform. Again, the very first sentence is the most important. This sentence must have the most impact. The best way to optimize such an approach is to use the journalist's inverted pyramid, even in scientific and technical writing. The following sentences, then subsequent paragraphs help to expand and explain in ever-growing detail what was initially communicated in the lead so that the most important items are easily gleaned within the first few seconds of reading. Again, most humans read documents (whether from a news source or in a scientific literature database) in such a manner anyway, similar to how our ancestors gained insights into their surroundings while on a hunt. Instead of fighting against human nature, scientific writers should embrace this psychological disposition and learn how to utilize it to the utmost advantage. Third person, present tense, active voice, omniscient viewpoint should be maintained in order to keep the information as objective as possible while still maintaining clear, concise, and correct structure. While the use of "we," "our," and "us" is technically correct and can be utilized, any sentence with first-person plural pronouns can be easily rewritten more succinctly and clearly without them.

Complete Science Communication: A Guide to Connecting with Scientists, Journalists and the Public
By Ryan C. Fortenberry
© Ryan C. Fortenberry 2019
Published by the Royal Society of Chemistry, www.rsc.org

3.2   Writing Abstracts

In truth, standard scientific writing has been employing the journalism approach for generations without even calling it such. This is the essence of the abstract. This short synopsis contains all of the high points of the discussion. It (1) sufficiently motivates the work, (2) provides a mention of the techniques employed to generate the new knowledge, (3) states the most important findings, and (4) gives a synthesis statement to describe why anyone should care based on what is reported and how it relates to the literature or an ongoing problem. The best abstracts contain a catchy lead sentence that doubles as a thesis addressing as many of these points as possible. This is one reason why the journalism chapter comes before this one. Then, the abstract has two to four (the more concise, the better) sentences addressing four points. These four points are leads for the typical four major sections of a scientific article: Introduction, Methods, Results, and Conclusions.

Some abstracts will have additional information after these four basic areas are addressed. However, this should only be done if the four major areas are fully described in succinct fashion. Trailing material in an abstract is quite acceptable as in most journalistic styles, but this should only be done if the four necessary pieces have some completeness to them. Again, the most important information in the abstract must come in the first, or first few, sentences.

The most important thing about an abstract is that it is concise. Some journals will allow abstracts to be as much as 300 words. This is way too much and is simply an artifact of historical length and not of practical, modern use. Shorter, high-impact articles like letters or communications have limited abstracts to 150 words. In truth, the best abstracts, regardless of the length of the corresponding paper, should be 150 words or less. Most readers will, just like in journalism, make it through the first sentence, but then, they start to pick out the "useful" information even in the abstract. The most readable abstracts are short and concise. They are dense but readable. No one is ever upset about getting as much information in 100 words as they would in 250. The author has lived this research and understandably wants the reader to love it (or hate it) as much as he or she has. In spite of such desires, a reader never will and never can. Hence, the shorter the material, the more likely the reader will appreciate and utilize the research.

One of the emerging trends in abstracts is the concept of the graphical abstract, also called the table of contents image, depending upon the journal. This is an image that contains catchy information about the paper's topic. In many ways this image is a visual lead. The journals do not want a regurgitated figure or really any text. This should be an image that has been developed in order to communicate the information present in the article. In so doing, the visual processing portion of the reader's brain can switch on, and the linguistic portion can turn off. Pictures truly are worth 1000 words (see the next chapter, especially, for this discussion) so a visual abstract can increase the communicative ability of the blurb in the table of contents in the journal five- or even sixfold if the visual abstract is good. These images are usually small, often less than 5 cm, but should still stand out in contrast to the text or plain background of the page. The best graphical abstracts catch the eye but do not overstimulate. They further describe the work done in visual terms and also lead to lots of questions about the research. Employing the techniques of making good visual slides (again, next chapter) will make the graphical abstract stand out. However, sometimes a simple reaction scheme is all that is necessary to serve as a graphical abstract.

3.2.1 Sample Abstracts

Abstract from: C. J. Stephan and R. C. Fortenberry, The interstellar formation and spectra of the noble gas, proton-bound $HeHHe^+$, $HeHNe^+$, & $HeHAr^+$ complexes, *Mon. Not. R. Astron. Soc.*, 2017, **469**, 339.

The sheer interstellar abundance of helium makes any bound molecules or complexes containing it of potential interest for astrophysical observation. This work utilizes high-level and trusted quantum chemical techniques to predict the rotational, vibrational, and rovibrational traits of $HeHHe^+$, $HeHNe^+$, and $HeHAr^+$. The first two are shown to be strongly bound, while $HeHAr^+$ is shown to be more of a van der Waals complex of argonium with a helium atom. In any case, the formation of $HeHHe^+$ through reactions of HeH^+ with HeH_3^+ is exothermic. $HeHHe^+$ exhibits the quintessentially bright proton-shuttle motion present in all proton-bound complexes in the 7.4 micron range making it a possible target for telescopic observation at the mid-IR/far-IR crossover point and a possible tracer for the as-of-yet unobserved helium hydride cation. Furthermore, a similar mode in $HeHNe^+$ can be observed to the blue of this close to 6.9 microns. The brightest mode of $HeHAr^+$ is dimmed due the reduced interaction of the helium atom with the central proton, but this fundamental frequency can be found slightly to the red of the Ar-H stretch in the astrophysically detected argonium cation.

Note that the first sentence sets the stage for the entire rest of the abstract. It comprises aspects of both lead and Introduction. The next sentence gives a brief but adequate description of the experimental (computational in this case) procedure sufficient for a Methods discussion. This is followed by the most important Result, that these structures are, in fact, real molecules. The second most significant Result, where in the infrared this molecule could be observed, is married to the Conclusions statement. Then, there are several statements of other Results in decreasing order of signficance. These trail off in standard journalistic style.

Abstract from: R. C. Fortenberry, T. J. Lee and J. P. Layfied, The failure of correlation to describe carbon=carbon bonding in out-of-plane bends, *J. Chem. Phys.*, 2017, **147**, 221101.

Carbon–carbon multiply bonded systems are improperly described with standard, wave function-based correlation methods and Gaussian one-particle basis sets, implying that thermochemical, spectroscopic, and potential energy surface computations are consistently erroneous. For computations of vibrational modes, the out-of-plane bends can be reported as imaginary at worst or simply too low at best. Utilizing the simplest of aromatic structures (cyclopropenylidene) and various levels of theory, this work diagnoses this known behavior as a combined one-particle and n-particle basis set effect for the first time. In essence, standard carbon basis sets do not describe equally well sp, sp^2, and sp^3 hybridized orbitals, and this effect is exacerbated post Hartree–Fock by correlation methods. The latter allow for occupation of the π and π^* orbitals in the expanded wave function that combine with the hydrogen s orbitals. As a result, the improperly described space is non-physically stabilized by post-Hartree–Fock correlation. This represents a fundamental problem in wavefunction theory for describing carbon.

Again, a statement that goes directly to the heart of the matter leads off the abstract in order to get attention and draw it directly to the matter at hand. A larger statment of background is given next by way of Introduction. The Methods are stated in a single, sufficient, but succinct sentence. (Yes, that is a purposeful alliteration; it makes the meaning more memorable.) Then, the Conclusions are actually stated. The Results back up the Conclusions in multiple statements, again, of decreasing significance.

Abstract and graphical abstract from: R. C. Fortenberry, R. Thackston, J. S. Francisco and T. J. Lee, Toward the laboratory identification of the not-so-simple NS_2 neutral and anion isomers, *J. Chem. Phys.*, 2017, **147**, 074303.

The NS_2 radical is a simple arrangement of atoms with a complex electronic structure. This molecule was first reported by Lester Andrew's group (*J. Am. Chem. Soc.* **114**, 83 (1992)) through Ar matrix isolation experiments. In the quarter century since this seminal work was published, nearly nothing has been reported about nitrogen disulfide even though NS_2 is isoelectronic with the common NO_2. The present study aims to shed new insight into possible challenges with the characterization of this radical. No less than three potential energy surfaces all intersect in the C_{2v} region of the SNS radical isomer. A type-C Renner–Teller molecule is present for the linear $^2\Pi_u$ state where the potential energy surface is fully contained within the 2.05 kcal/mol^{-1} lower energy $\tilde{X}\ ^2A_1$ state. A C_{2v}, $1\ ^2B_1$ state is present in this same region, but a double excitation is required to access this state from the $\tilde{X}\ ^2A_1$ state of SNS. Additionally, a $1\ ^2A'$ NSS isomer is also present but with notable differences in the geometry from the global minimum. Consequently, the rovibronic spectrum of these NS_2 isomers is quite complicated. While

Figure 3.1 Graphical abstract sample regarding multiple states of NS_2 and the difficulty in determining them.

the present theory and previous Ar matrix experiments agree well on isotopic shifts, they differ notably for the absolute fundamental vibrational frequency transitions. These differences are likely a combination of matrix shifts and issues associated with the neglect of non-adiabatic coupling in the computations. In either case, it is clear that high-resolution gas-phase experimental observations will be complicated to sort out. The present computations should aid in their analysis.

While the *Journal of Chemical Physics* does not actually publish a graphical abstract, the image given in Figure 3.1 was developed as such, for illustrative purposes. In any case, the graphical abstract communicates how different potential energy surfaces, which are often depicted as ridges and valleys, intersect in a seeming mountain range where the actual structure of the molecule is hiding from sight. The regular abstract follows the standard style as those above but also has the accompanying image. Such an image evokes enough information for questions to be formed by the interested audience. Even if the audience is not ultimately concerned with the actual study, the image draws enough attention in its novel portrayal to garner additional attention at least for an initial analysis by the reader.

3.3   The Four Major Parts of a Scientific Article

While *the most* important sentence is always the first of every document, the first sentence beginning each section should be the most important of each section. The rest of the text in each section supports the lead, again, just like in journalistic writing. The typical sections in a scientific paper include: Introduction, Methods, Results, and Conclusions. This structure exists in order to allow the reader to find the necessary information easily. Many researchers will already know the background sufficiently and can skip the Introduction and go straight to the Methods section. Some will find the Introduction highly useful for broadening their knowledge in the field, but the Methods may be beyond their need to grasp. Most, however, will skim or skip through that material feeling as though they gleaned enough from the Abstract and go straight to the data in the tables and figures and/or read the Conclusions. Hence, the second most important sentence in the entire document is the first sentence of the Conclusions. That is the proper synthesis/highlights sentence that most readers will need. It should mirror the most important sentence of the entire document:

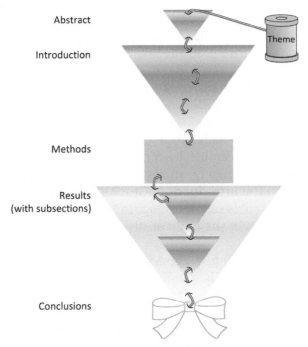

Abstract

Introduction

Methods

Results
(with subsections)

Conclusions

Figure 3.2　The structure of the overall scientific paper utilizing journalistic techniques for each section.

the lead of the abstract. A visual depiction for the inclusion of journalistic techniques in the overall structure of the technical paper is given in Figure 3.2.

When writing a paper, typically, the Methods are the first section written, especially for new scientists. These are basically clear instructions for someone else to do the research but told in a narrative, not journalistic, fashion save for a lead (discussed below). This is the *only* section written this way. Any student or scientist who struggles with writing should be able to compose a prosaic form of steps that describe exactly what was done in the science. Starting from this point is also less daunting since most practitioners will have lived these steps for the previous weeks, months, or even years. Writing a lead may be a bit of a challenge, but a summative statement can be succinctly constructed if it contains the basic methodology type within. For instance, "This quantum chemical analysis relies upon coupled cluster theory to compute potential energy surfaces for the dissociation of the various ionic compounds in question." Such a sentence fully describes the research procedure. The finer details are implied to be contained within the rest of the Methods section.

Just like with journalistic pieces, the third person, active voice, omniscient viewpoint must be maintained even in the Methods. Present tense can be a bit tricky here since it may seem weird to say that "50.0 mL of solution are titrated into a beaker" when this actual titration happened months ago. However, the fact that the reader is experiencing this from the text in the present mandates that the present tense should be retained. Also, all abbreviations and acronyms must be defined upon first use. This is typically in the Methods but can be in the Introduction. Note that the abstract and body of the paper are viewed as separate documents. Hence, all abbreviations should be redefined in the body of the text even if first mentioned in the abstract.

Beyond the first sentence, the Methods are often dry. They do not need to be flowery or really that dense. The Methods need to be clear. Again and most importantly, the Methods need to be clear. One of the biggest problems in science today is the reproducibility of research. Beyond unethical practices, most of this is a result of poorly written Methods sections. The Methods exist so that someone else may come along and reproduce the study. Such a follow-up is not to generate the same results and say, "Yes, Prof. Wilson did perform her study correctly." The real reason for reproducibility is that Prof. Wilson can later build upon the work done in order to take it a step beyond. The next step on the stairs can only be accessed if the present step is solid. Hence, the Methods must be clear as well as complete. Every bit of information done must go into the Methods section.

Once the full Methods are described, then the numbers produced will be discussed in the Results. The first thing to be done is to make all of the pertinent figures and tables. Then, write the tables; write the figures. The prose of the article *does not* serve to regurgitate the data on those tables and figures; the text augments such data. The tables and figures should largely speak for themselves. The reader typically will gloss over walls of text but will ponder visual data much more intently. This should not be fought but must be understood in order to make scientific communication more effective. The prose, on the other hand, must serve to interpret the tables and figures, to provide synthesis of ideas between parts of the data depicted, and to address any issues with gaps, holes, or potentially misleading information. In truth, the Results really are the visual or tabular data. The text just makes them better developed.

In formulating how to write the Results, the author should, for his or her own pre-writing exercise, highlight or make note of the most important items in the tables or figures. Once this list is compiled, the author can then order these items in terms of significance with the most important items coming first. This serves as the outline for the Results section (the reader should know that "Results" also refers to the "Results and Discussion" combined section). Then, these items can be fleshed out into paragraphs in journalistic style with the most important ideas coming first. Sometimes this means that less important but more fundamental results need to be mentioned so that the most import findings can be reported, but they can be discussed in further detail later, again like in a journalistic piece.

Regardless, the paragraphs should then be connected. The two best ways to connect the paragraphs are through comparison and contrast of the previous point with the current one or by relating each item to the theme or motivation of the work like the thread in Figure 3.2. This keeps the discussion moving in a natural way. These points should not be bullet point paragraphs but should connect with one another as part of the larger narrative. Synthesis/conclusions paragraphs are not necessary or even useful in the Results since scientific articles really will be read, again, in much the same way as journalistic pieces. A good continuance of the discussion is much more important than a good concluding statement. The latter will be dealt with in the Conclusions anyway.

At this point, the Introduction can be composed. Some prefer to do this before the Results in order to have a clear tie between the two. However, that is a matter of style and experience. The motivation for the study must be contained in the discussion present in the Results. Otherwise, the research stands as an uninhabited island. However, the motivation for the work may not be fully fleshed out until the study findings are produced (serendipity still exists), or the study may simply be a continuance straight from prior knowledge. In either preferred case of writing the Results or the Introduction first, the two must be massaged together through subsequent editing and drafting. The Introduction can be such a difficult beast that it will be discussed in its own section below.

Regardless of whatever the order is for the writing. The Conclusions must be written last. Most often, the Conclusions are viewed as a summary. They are *NOT*. They are both a synthesis and an evaluation, hence the bow in Figure 3.2. The pieces of the Introduction producing motivation must be combined with the Results to convince the reader that something new and useful building upon prior knowledge or an existing problem has been produced with proper application of the Methods. Furthermore, the Conclusions imply whether the study was useful and, if so, to whom the Results will apply. They are the final piece that holds the document together. And, yes, the first sentence of the Conclusions is probably the most important in the entire document besides the first sentence of the abstract. The abstract and the Conclusions may often seem to be the same thing. Occasionally, this author has even swapped the two out in papers. However, the biggest difference is that the abstract provides an overall synopsis of the four sections with a lead whereas the Conclusions make the reader feel as though the Results are simply the next step in a continuation of the material in the Introduction.

The one major exception to this structure comes in "letters" or "communications." These are typically short ($\sim < 2200$ words) and are of high significance to the field. Additionally, these may not be fully fleshed out ideas but are so potentially impactful that the community needs to know about them as soon as possible. In writing such pieces, the standard format is forgotten. The entire narrative flows more naturally so that it can be digested more quickly. These articles are not meant to be reference articles; they are the first foray into something truly novel. More journals are moving to this model for many article styles. Hence, these types of articles should really be treated as single 2000 word journalistic pieces where the scientist who did the study is writing the news article for his or her peers. This should be done with technical language, but such articles will be even more impactful by maintaining the journalistic style of having the most important information first and employing the larger inverted pyramid style. However, a concluding paragraph in such pieces would still be a good idea because some readers will still skip to the end to find the punchline.

The final bit in a peer-reviewed paper is the Acknowledgements section. This is most notably where the funding agencies are given credit for provision of funds necessary to do the work. Additionally, thanks should be given to those who helped with the paper but are not given credit as authors (due to simply giving advice but not engaging in the experimentation or writing) or to anyone who was otherwise helpful in the process. Marriage proposals have even been put in Acknowledgements before. Finally, there may be a statement for Conflicts of Interest, or explanations of any supplemental (additional online) information that could be useful but was too bulky or not significant enough for inclusion in the main body of the text.

Table 3.1 CCSD(T)/aug-cc-pV5Z formation/destruction energetics.

Reactions	kcal/mol^{-1}	eV
$He + HeH^+ \rightarrow HeHHe^+$	−13.2	−0.57
$2He + H^+ \rightarrow HeHHe^+$	−60.3	−2.62
$2HeH^+ \rightarrow HeHHe^+ + H^+$	33.8	1.47
$HeH^+ + HeH_3^+ \rightarrow HeHHe^+ + H_3^+$	−11.9	−0.52
$He + H_3^+ \rightarrow HeH_3^+$	−1.3	−0.06
$He + NeH^+ \rightarrow HeHNe^+$	−11.9	−0.52
$Ne + HeH^+ \rightarrow HeHNe^+$	−21.9	−0.95
$HeH^+ + NeH^+ \rightarrow HeHNe^+ + H^+$	35.2	1.52
$NeH^+ + HeH_3^+ \rightarrow HeHNe^+ + H_3^+$	−10.6	−0.46
$HeH^+ + NeH_3^+ \rightarrow HeHNe^+ + H_3^+$	−15.4	−0.67
$Ne + H_3^+ \rightarrow NeH_3^+$	−6.5	−0.28
$He + ArH^+ \rightarrow HeHAr^+$	−2.1	−0.09
$Ar + HeH^+ \rightarrow HeHAr^+$	−49.0	−2.13
$HeH^+ + ArH^+ \rightarrow HeHAr^+ + H^+$	45.0	1.95
$ArH^+ + HeH_3^+ \rightarrow HeHAr^+ + H_3^+$	−0.8	−0.03
$HeH^+ + ArH_3^+ \rightarrow HeHAr^+ + H_3^+$	−40.0	−1.73
$Ar + H_3^+ \rightarrow ArH_3^+$	−9.1	−0.39

3.3.1 Sample Results Section

The first portion of the Results from: C. J. Stephan and R. C. Fortenberry, The interstellar formation and spectra of the noble gas, proton-bound HeHHe$^+$, HeHNe$^+$, & HeHAr$^+$ complexes, *Mon. Not. R. Astron. Soc.*, 2017, **469**, 339.

> The interstellar presence for any of these proton-bound complexes can be established by observing the necessary rotational or vibrational transitions. However, the possible formation of these species must be established in order for such searches to be deemed even plausible before observation can begin. In light of such, high-level computations at the CCSD(T)/aug-cc-pV5Z level are performed here in order to determine if these proton-bound complexes can form exothermically in the gas-phase. Similar and previous analysis[46] shows that ArHAr$^+$ is a likely interstellar proton-bound complex created from the reaction of ArH$^+$ with ArH$_3^+$. Table 3.1 contains these data for HeHHe$^+$, HeHNe$^+$, and HeHAr$^+$ in the standard chemistry unit of kcal/mol^{-1} as well as in eV. The complexes lie to the right of the reaction arrow indicating that negative energies are desirously exothermic in the gas-phase.
>
> HeHHe$^+$ is thermodynamically stable since removal of a single helium atom has a 0.57 eV energy cost. Removal of both helium atoms is over five times higher in energy. The collision of two

helium hydride cations will not form $HeHHe^+$ and a hydrogen atom. However and like with $ArHAr^+$, the reaction of helium hydride with helium trihydride produces $HeHHe^+$ with an excess of 0.52 eV of energy. The formation of HeH_3^+ is a matter of further discussion, but HeH_3^+ is stable requiring a small but non-negligible amount of energy to dissociate the helium atom.

$HeHNe^+$ is also thermodynamically stable, but removal of the helium atom (–0.52 eV) is nearly half as energetically costly as removal of the neon atom at –0.95 eV. This is likely actually the strongest neon bond produced thus far besides neonium itself. The neon bonding in $NeOH^+$ is –0.53 eV, $NeCCH^+$ is –0.93 eV, and $NeHNe^+$ –0.87 eV. Helonium reacting with NeH_3^+ favors creation of $HeHNe^+$ and the ubiquitous H_3^+ with –0.67 eV. Reacting neonium with He_3^+ is also exothermic but less favorable at –0.46 eV.

$HeHAr^+$ is much more dichotomous. The helium is not well-bound (–0.09 eV), but the argon atom very much is (–2.13 eV). This trend is carried through the other formation and destruction reactions listed in Table 3.1. However, ArH_3^+ has the strongest Ng–H bond of all the noble gas-trihydride cations and is hypothesized to exist in the ISM[17,47]. Reacting this species with the almost guaranteed interstellar HeH^+ produces $HeHAr^+$ and –1.73 eV of energy.

Hence, $HeHAr^+$ is the most energetically-favored product, but interstellar abundances still point to $HeHHe^+$ as the most likely to be observed. In any case, only spectroscopic observation in the ISM can determine the presence for any of these compounds. The rovibrational computed data for each complex (not given here) should be able to assist in the laboratory or even interstellar observation of these noble gas, molecular cations.

3.4 Writing the Introduction and Utilizing the Scientific Literature

As always, the first, lead sentence of the Introduction sets the tone for the entire document thereafter. Hence, it should contain as much useful information for the setup of the piece as possible. The best way to construct this is to utilize the 5 "Ws" of journalism, just like in a journalistic article lead. The difference is that the lead for the Introduction of a scientific research article is a statement of the problem to be addressed and not a statement of the resolution; the latter is the essence of the lead for the Conclusions. Such a sentence is discussed previously in this chapter. Again, this Introduction lead sets up the rest of the section. The Introduction largely should be written like a 500 word journalistic piece. After the lead, the rest of the first paragraph develops the lead and fills in the gaps. Then, the rest of the ~400 words of the Introduction should further expand on the first paragraph, just like in the news pieces discussed in the previous chapter.

A key difference between scientific writing and journalism comes in sources. The background information from the scientific literature should be discussed concisely in order to motivate the work. The narrative will point towards the ultimate findings of the research described later, but it will not deliver *any* of those. That is the purpose of the Results. Formally and scientifically, the *only* sources that are pertinent are those that come from the peer-reviewed literature. Granted, these may span several academic silos in our modern age of interdisciplinary research, but peer-reviewed articles are the only acceptable sources.

A simple way to construct the Introduction begins with the gathering of sources (discussed in the following paragraphs) and the author familiarizing himself or herself with them. At that point, the sources should be listed out using some type of shorthand. This could be the title of the paper, the author, or something unique to each paper. This author uses a key that includes the first (or major) author's last name, the two-digit year, and some descriptor. For instance, "Hogness25HeH+" refers to the first experimental study of HeH^+ and is ref. 1 in the sample Introduction given later. These cite keys should be put in a list in order to help the author create a critical mass of sources necessary for the writing.

Then, these cite keys can be arranged in the list based on their importance. This creates a working outline and ensures that the author includes the necessary references. Next, the cite keys should turn into brief descriptions/summaries/synopses of the work. Now, the flesh is being put on to the bones of the manuscript. Once these descriptions (sort of like an annotated bibliography) are present, they can be stitched together and massaged through in order to create a functional Introduction. However, it all starts with having the proper list of references.

The best place to find reference sources is in a database. There are several including services like EBSCO Host, SciFinder, Chemical Abstract Services, and even publishers like the Royal Society of Chemistry. The key is remembering that the scientific literature is a web with the interconnects being the subsequent papers citing, and the previous papers cited within, a given article. A few key words are typed into a database search while looking for sources. Then, one article jumps out (or is already known, potentially thrust at you by a research mentor) as being an excellent starting point for your research. This is the sentinel article.

Fleshing out the scientific literature really is easy at this point. It may be time-consuming but not mentally taxing. Find the pertinent articles that the sentinel work cites. Find the ones that those cite. Then, find the articles that cite the sentinel work and the ones that cite or are cited by that one. Eventually, a critical mass will start to appear, and not every work cited by every paper in this web will be useful. However, this body of literature is the basis for a story to tell. The proper number of references is "enough" and not some set amount. However, a general rule of thumb is that there should be one reference, on the average and at least, for every 1.5 sentences in the Introduction.

Once this body of literature is on-hand and has been processed, then, the author can take these pieces and put them together in a journalistic narrative as described above. The lead is the foundation for the Introduction, and the cited references are the framed-out walls of the structure. Without these pieces, the building will collapse. Then, the surfaces of the walls, *i.e.* the author's words used to make these pieces coalesce, can be included.

For instance, seemingly disparate sources relating to spectral analysis may be necessary, but sources related to environmental degradation could also be pertinent to the given work. These different but essential concepts have to be fleshed out and done so together in order to motivate an example study on using remote sensing for monitoring smog levels in urban environments. Regardless, every statement or essence must be backed up by these sources. Quotations should be kept to a minimum (or be non-existent) unless a real, true nugget of text warrants such an inclusion. Many previous studies have built upon one another, and their essences can be distilled down into a single thought constructed by the author. However, all of these sources must be cited even if cited together.

The end of the Introduction should largely trail off just as a journalistic piece. As the discussion narrows and goes from general to being specific to the research, the ultimate narrowing of the Introduction ultimately leads to the last paragraph, or even sentence, of the Introduction. This final thought is often the primer of what the present study will actually be doing in order to add to the literature previously discussed in this section. Statements about the arrangement of the result of the paper are useless wastes of text and should never be included. They add nothing to the Introduction.

At any rate, few readers will make it much into the Introduction before skipping past the first paragraph and diving into the Methods (or going straight to the Conclusions) which is completely fine and within human nature even if it is aggravating for authors. As such, most Introductions are far too long. The higher the citation-to-word ratio, the better. Hence, "Concise" of the three "C's" of science communication also works extremely well in writing papers, especially the Introduction. Most reviewers are tasked with, among other things, offering suggestions to the author or the editor about how the text can be reduced. Keeping the Introduction to the $\sim$500 word journalistic length is usually a good rule of thumb.

3.4.1 Introduction Example

From: C. J. Stephan and R. C. Fortenberry, The interstellar formation and spectra of the noble gas, proton-bound $HeHHe^+$, $HeHNe^+$, & $HeHAr^+$ complexes, *Mon. Not. R. Astron. Soc.*, 2017, **469**, 339.

This sample Introduction is taken straight from the above source. It immediately highlights at the very beginning that most of the Universe is not made of most atoms associated with chemistry. This point is explained in the remaining paragraph with details given about more noble gas chemistry, especially that of the ArH^+ cation. In so doing, it sets the stage for gaining the reader's interest and keeping it going by countering many common beliefs about noble gases and their supposed lack of chemistry. Then, the discussion narrows to the types of noble gas molecules present for the three lightest noble gas atoms and then narrows further to the types of molecule to be studied in this work, proton-bound complexes. Finally, the methods utilized in the work (but not the actual details which come in the Methods section) are included and references for their benchmarks and behaviors are given as well as possible applications for the present work. This format works very well to attract readers who can break off their read as the Introduction progresses if they are no longer interested in the finer details of the study. However, it gives them the chance to get the broad picture at the very beginning, just like journalism.

Please note that the style of the publications was originally placed in the journal *Mon. Not. R. Astron. Soc.* which differs from the standard Royal Society of Chemistry style for the references and the journal citations that is utilized here. However, the general premise remains intact. The reference list always comes at the end of the document but will be placed at the end of this section for the sake of the present discussion.

Helium and hydrogen make up nearly all of the observable matter in the universe leaving chemists to squabble over the remaining scraps. These scraps are what compose the planets, our bodies, and most other things engineered by human beings. Nearly all other processes depend upon atoms much more interesting than the first two on the periodic table. Even so, helium and hydrogen can engage in chemistry with one another almost certainly combining to make HeH^{+} [1]. This cation should be produced in detectable amounts if for no other reason than the sheer abundance of the constituents in the interstellar medium (ISM)[2]. However, such an interstellar observation of this diatomic cation has yet to be reported in the literature. It was the analogous ArH^{+} that has been observed according to various astronomical sources[3-6] making the argonium and not helonium (helium hydride) cation the first noble gas molecule detected in nature. The smaller and more abundant helium and even neon hydride cations have not been observed, yet.

The chemistry of helium is likely the least voluminous for any of the elements between hydrogen and iron even in controlled laboratory conditions. However, helium will make complexes and form some bonds. Helium cationic clusters have been predicted, $He_mH_n^{+}$ clusters have been synthesized, dication complexes observed, and even hydrogen-like replacement structures analyzed[7-11]. In all cases, the issue is that the helium cation binding in any of these complexes is relatively weak, making long-lifetime molecules and high enough abundances for observable interstellar spectra of such chemical combinations quite unlikely.

Like unto helium, neon is reluctant to form bonds. There is little surprise here due to the high ionization potentials and relatively poor polarizabilities in these smallest of noble gas compounds[12-17]. Neonium (NeH^{+}) has been well-characterized[18-22], but it has yet to be conclusively observed in any astrophysical environment. While the reaction of Ar^{+} with ubiquitous hydrogen gas leads to ArH^{+} and hydrogen atoms in the ISM, the analogous reaction with neon will initially lead to neutral neon atoms and ionized hydrogen gas[23]. More complicated neon structures beyond NeH^{+} have been proposed and even synthesized, but few have bond strengths in the covalent range[7,24]. Notable exceptions include $NeOH^{+}$ and $NeCCH^{+}$ recently characterized theoretically at high level[25,26], but these are several factors less stable than their argon counterparts. With helium and neon being so very abundant in the ISM, neon even more so than nitrogen[27], molecules containing these atoms may still be awaiting detection.

These molecules in waiting could be proton-bound complexes. These structures involve a mostly bare proton situated between two other atoms or molecules where the mutual attraction of the ligands to the proton creates fairly strong interactions. These structures are of additional significance to astrochemistry and astrophysical observation due to their proton "rattle" or "shuttle" motion. In such vibrational modes, basically only the proton moves. Hence, little of the mass but nearly all of the charge is moving, creating an immense change in dipole moment. As a result, such vibrational transitions are incredibly strong absorbers/emitters meaning that small column densities of materials are required to create observable spectral features. $OCHCO^+$, $NNHNN^+$, and the heteromolecular combinations have been analyzed recently showing that these bright vibrational modes can be found from the near- to mid- and even far-IR wavelengths[28-34] making them tantalizing targets for the epoch for growth in IR telescopic power in which we are currently in the midst.

Proton-bound complexes of noble gases have been known for some time. In fact, the simplest, $HeHHe^+$, was noted for its relatively strong bonds nearly 35 years ago[35]. However, a complete and reliable set of rovibrational spectroscopic data have yet to be produced for this simple system while other insights into its nature have been explored theoretically[36-38]. Other data for related noble gas molecules have been produced including those with helium[21,39-44], but full spectral charactization is still lacking for most of these structures.

Very recently, the vibrational spectra of Ar_nH^+ complexes were characterized experimentally including $ArHAr^+$ [45] with its bright proton motion at 10.11 microns with dissociation not occurring until 1.74 microns (or 0.711 eV). Simultaneously, theoretical work on this complex produced a very similar dissociation energy (0.719 eV) and comparable vibrational frequencies[46]. The $NeHNe^+$ and $NeHAr^+$ complexes were also analyzed. Fortenberry[46] showed that the $NeHNe^+$ complex is actually more strongly bound than $ArHAr^+$ indicating that, for once, the neon bonds are actually stronger than the more polarizable argon bonds in a cation. The $NeHNe^+$ dissociation is higher at 0.867 eV, but interstellar synthesis of

NeHNe$^+$ is most likely in the gas phase from reactions of NeH$^+$ with NeH$_3^+$. Again, the former has yet to be observed, and the latter is only weakly bound[47]. ArHAr$^+$ is also favorably created from ArH$^+$ and ArH$_3^+$ where the former is, again, known in the ISM and the latter is a viable interstellar candidate[14,17,47]. Additionally, ArHAr$^+$ has a much brighter and longer wavelength proton shuttle motion making it more likely to be observed in the ISM[46].

Consequently, the question lingers as to whether proton-bound complexes involving the abundant helium atom are viable interstellar detection candidates. Furthermore, combinations of noble gas atoms in such complexes with helium are known to be fairly stable[41,44] and those with other noble gas atoms have been classified at high-level with good comparison to experiment[46]. As a result, this work will employ the same methodology as that utilized previously on proton-bound complexes[32,33,46] where comparison in other molecules to gas phase experimental results has provided exceptional accuracy on the order of 0.01 micron accuracy for vibrational features and 30 MHz for rotational constants[48-61]. These data will be useful for the spectral characterization of such molecules in the ISM with current and upcoming ground- and space-based telescopes such as the upcoming James Webb Space Telescope. Potentially expanding the noble gas molecular budget of the ISM will grow our understanding of interstellar chemical bonding and provide novel chemical pathways for these so-called "inert" and noble gases.

References

1. T. R. Hogness and E. G. Lunn, *Phys. Rev.*, 1925, **26**, 44–55.
2. W. Roberge and A. Dalgarno, *Astrophys. J.*, 1982, **255**, 489–496.
3. M. J. Barlow, B. M. Swinyard, P. J. Owen, J. Cernicharo, H. L. Gomez, R. J. Ivison, O. Krause, T. L. Lim, M. Matsuura, S. Miller, G. Olofsson and E. T. Polehampton, *Science*, 2013, **342**, 1343–1345.
4. P. Schilke, D. A. Neufeld, H. S. P. Müller, C. Comito, E. A. Bergin, D. C. Lis, M. Gerin, J. H. Black, M. Wolfire, N. Indriolo, J. C. Pearson, K. M. Menten, B. Winkel, A. Sánchez-Monge, T. Möller, B. Godard and E. Falgarone, *Astron. Astrophys.*, 2014, **566**, A29.
5. E. Roueff, A. B. Alekseyev and J. L. Bourlot, *Astron. Astrophys.*, 2014, **566**, A30.
6. D. A. Neufeld and M. G. Wolfire, *Astrophys. J.*, 2016, **826**, 183.
7. G. Frenking and D. Cremer, *Struct. Bonding.*, 1990, **73**, 17–95.
8. D. Roth, O. Dopfer and J. P. Maier, *Phys. Chem. Chem. Phys.*, 2001, **3**, 2400–2410.
9. F. Grandinetti, *Int. J. Mass Spectrom.*, 2004, **237**, 243–267.
10. I. Savic, D. Gerlich, O. Asvany, P. Jusko and S. Schlemmer, *Mol. Phys.*, 2015, **113**, 2320–2332.
11. E. Zicler, M.-C. Bacchus-Montabonel, F. Pauzat, P. Chaquin and Y. Ellinger, *J. Chem. Phys.*, 2016, **144**, 111103.

12. P. R. Taylor, T. J. Lee, J. E. Rice and J. Almlöf, *Chem. Phys. Lett.*, 1989, **163**, 359–365.
13. J. E. Rice, P. R. Taylor, T. J. Lee and J. Almlöf, *J. Chem. Phys.*, 1991, **94**, 4972–4979.
14. F. Pauzat and Y. Ellinger, *Planet. Space Sci.*, 2005, **53**, 1389.
15. F. Pauzat and Y. Ellinger, *J. Chem. Phys.*, 2007, **127**, 014308.
16. F. Pauzat, Y. Ellinger, J. Pilmè and O. Mousis, *J. Chem. Phys.*, 2009, **130**, 174313.
17. F. Pauzat, Y. Ellinger, O. Mousis, M. A. Dib and O. Ozgurel, *Astrophys. J.*, 2013, **777**, 29.
18. R. Ram, P. Bernath and J. Brault, *J. Mol. Spectrosc.*, 1985, **113**, 451–457.
19. F. Matsushima, Y. Ohtaki, O. Torige and K. Takagi, *J. Chem. Phys.*, 1998, **109**, 2242–2245.
20. P. Gamallo, F. Huarte-Larrañaga and M. González, *J. Phys. Chem. A*, 2013, **117**, 5393–5400.
21. D. Koner, L. Barrios, T. Gonzalez-Lezana and A. N. Panda, *J. Chem. Phys.*, 2016, **144**, 034303.
22. J. A. Coxon and P. G. Hajigeorgiou, *J. Mol. Spectrosc.*, 2016, **330**, 63–71.
23. R. A. Theis, W. J. Morgan and R. C. Fortenberry, *Mon. Not. R. Astron. Soc.*, 2015, **446**, 195–204.
24. F. Grandinetti, *Eur. J. Mass Spectrom.*, 2011, **17**, 423–463.
25. R. A. Theis and R. C. Fortenberry, *Mol. Astrophys.*, 2016, **2**, 18–24.
26. C. M. Novak and R. C. Fortenberry, *Phys. Chem. Chem. Phys.*, 2017, **19**, 5230–5238.
27. B. D. Savage and K. R. Sembach, *Annu. Rev. Astron. Astrophys.*, 1996, **34**, 279–329.
28. K. Terrill and D. J. Nesbitt, *Phys. Chem. Chem. Phys.*, 2010, **12**, 8311–8322.
29. C. E. Cotton, J. S. Francisco, R. Linguerri and A. O. Mitrushchenkov, *J. Chem. Phys.*, 2012, **136**, 184307.
30. R. C. Fortenberry, Q. Yu, J. S. Mancini, J. M. Bowman, T. J. Lee, T. D. Crawford, W. F. Klemperer and J. S. Francisco, *J. Chem. Phys.*, 2015, **143**, 071102.
31. Q. Yu, J. M. Bowman, R. C. Fortenberry, J. S. Mancini, T. J. Lee, T. D. Crawford, W. Klemperer and J. S. Francisco, *J. Phys. Chem. A*, 2015, **119**, 11623–11631.
32. R. C. Fortenberry, T. J. Lee and J. S. Francisco, *Astrophys. J.*, 2016, **819**, 141.
33. R. C. Fortenberry, T. J. Lee and J. S. Francisco, *J. Phys. Chem. A*, 2016, **120**, 7745–7752.
34. S. Begum and R. Subramanian, *J. Mol. Model.*, 2016, **22**, 6.
35. C. E. Dykstra, *J. Mol. Struct.*, 1983, **12**, 131–138.
36. I. Baccarelli, F. A. Gianturco and F. Schneider, *J. Phys. Chem. A*, 1997, **101**, 6054–6062.
37. A. N. Panda and N. Sathyamurthy, *J. Phys. Chem. A*, 2003, **107**, 7125–7131.
38. P. Bartl, C. Leidlmair, S. Denifl, P. Scheier and O. Echt, *Chem. Phys. Chem.*, 2013, **14**, 227–232.
39. T. D. Fridgen and J. M. Parnis, *J. Chem. Phys.*, 1998, **109**, 2162–2168.
40. J. Lundell, M. Pettersson and M. Rasanen, *Phys. Chem. Chem. Phys.*, 1999, **1**, 181–194.
41. D. Koner, A. Vats, M. Vashishta and A. N. Panda, *Comput. and Theor. Chem.*, 2012, **1000**, 19–25.
42. D. Koner, L. Barrios, T. Gonzalez-Lezana and A. N. Panda, *J. Chem. Phys.*, 2014, **141**, 114302.
43. S. Borocci, M. Giordani and F. Grandinetti, *J. Phys. Chem. A*, 2015, **119**, 6528–6541.
44. S. J. Grabowski, J. M. Ugalde, D. M. Andrada and G. Frenking, *Chem. Euro. J.*, 2016, **22**, 11317–11328.
45. D. C. McDonald II, D. T. Mauney, D. Leicht, J. H. Marks, J. A. Tan, J.-L. Kuo and M. A. Duncan, *J. Chem. Phys.*, 2016, **145**, 231101.
46. R. C. Fortenberry, *ACS Earth Space Chem.*, 2017, **1**, 60–69.
47. R. A. Theis and R. C. Fortenberry, *J. Phys. Chem. A*, 2015, **119**, 4915–4922.
48. X. Huang and T. J. Lee, *J. Chem. Phys.*, 2008, **129**, 044312.
49. X. Huang and T. J. Lee, *J. Chem. Phys.*, 2009, **131**, 104301.

50. X. Huang, P. R. Taylor and T. J. Lee, *J. Phys. Chem. A*, 2011, **115**, 5005–5016.
51. D. Zhao, K. D. Doney and H. Linnartz, *Astrophys. J. Lett.*, 2014, **791**, L28.
52. R. C. Fortenberry, X. Huang, J. S. Francisco, T. D. Crawford and T. J. Lee, *J. Chem. Phys.*, 2011, **135**, 134301.
53. R. C. Fortenberry, X. Huang, J. S. Francisco, T. D. Crawford and T. J. Lee, *J. Chem. Phys.*, 2011, **135**, 214303.
54. R. C. Fortenberry, X. Huang, J. S. Francisco, T. D. Crawford and T. J. Lee, *J. Chem. Phys.*, 2012, **136**, 234309.
55. X. Huang, R. C. Fortenberry and T. J. Lee, *J. Chem. Phys.*, 2013, **139**, 084313.
56. X. Huang, R. C. Fortenberry and T. J. Lee, *Astrophys. J. Lett.*, 2013, **768**, 25.
57. R. C. Fortenberry, X. Huang, T. D. Crawford and T. J. Lee, *Astrophys. J.*, 2013, **772**, 39.
58. R. C. Fortenberry, X. Huang, T. D. Crawford and T. J. Lee, *J. Phys. Chem. A*, 2014, **118**, 7034–7043.
59. R. C. Fortenberry, T. J. Lee and H. S. P. Müller, *Mol. Astrophys.*, 2015, **1**, 13–19.
60. M. J. R. Kitchens and R. C. Fortenberry, *Chem. Phys.*, 2016, **472**, 119–127.
61. R. C. Fortenberry, E. Roueff and T. J. Lee, *Chem. Phys. Lett.*, 2016, **650**, 126–129.

3.4.2 Sample Conclusions

From: C. J. Stephan and R. C. Fortenberry, The interstellar formation and spectra of the noble gas, proton-bound $HeHHe^+$, $HeHNe^+$, & $HeHAr^+$ complexes, *Mon. Not. R. Astron. Soc.*, 2017, **469**, 339.

This sample Conclusions section is taken straight from the above source, as well. The Conclusions immediately go straight at what the reader should find most useful. This is then further expanded in this paragraph. A next most important point is described in the second paragraph with more discussion. The first describes the bonding necessary to form a molecule while the second gives vibrational spectroscopic implications of the first point discussed in this Conclusions section. The vibrational signatures are applied to astrophysical observations per the theme of this paper, and the text gives some qualitative application of what the results can do for interpretation of astrochemical data in a larger sense.

> The most striking result from this study in light of the earlier work by ref. 46 is that neon and helium behave very similarly while argon does not. The intensities, fundamental vibrational frequencies, and even binding energies are not significantly changed when moving down the periodic table from helium to neon in such proton-bound complexes. Once argon is invoked, the chemistry changes fundamentally. This is likely due to the polarizability of argon and the energy proximity of the additional d orbitals close to argon's valence orbital occupation.

> In any case, the proton-bound complexes involving helium and neon ($HeHHe^+$, $HeHNe^+$, and $NeHNe^+$) give very intense proton shuttle motions in the range where the mid-IR becomes the far-IR.

The newest generation of space-based telescopes like the upcoming James Webb Space Telescope or even the Stratospheric Observatory for Infrared Spectroscopy can potentially be utilized to observe these vibrational frequencies. The $HeHNe^+$ dipole moment is also large, making this proton-bound complex a candidate for rotational observation with ground-based telescopes as has been the common practice for half a century. The $HeHAr^+$ complex appears to be readily formed from hypothesized interstellar species, the helonium and argon trihydride cations. $HeHAr^+$ is also rotationally active and has a bright fundamental vibrational frequency, but both of these are reduced relative to $HeHNe^+$.

3.5 Telling an Effective Story

The key to telling a story is to find a common thread and continually relate all parts of the narrative back to this as depicted in Figure 3.2. This thread should be first described as the problem to be answered in the first paragraph of the Introduction if not the lead sentence itself. Once this has been established, reminding the reader as to why the previous research helped to answer this question even if not fully; how the Methods are helping to answer this question; what the new data are and how they are giving new insights; and how these new data are providing as great as a paradigm shift or simply a further step in the ladder of knowledge, makes the text readable.

For example, writing a paper on the infrared spectral features of a peptide chain will get extremely dry and dull if the features from the tables are simply pointed out. However, directing the reader to how these spectral features fall into a certain range or influence the interpretation of old data from previous work in new ways is essential in making a paper useful and interesting to those who might care. Additionally, keeping this common thread going will keep the reader engaged in wanting to know more and will bring the manuscript together sensibly. Then, the Conclusions are a natural next progression even if the information in the Results, for instance, simply begins to close as is journalistic style.

The easiest way to conceptualize writing a scientific piece is to think of these parts as floats in a parade (with the sections of the paper as floats). The parade coordinator (or author as it were), can have a bird's eye view of the entire progression. Each float, transition, turn, and progression in the route is observed in its entirety. However, the parade-goer (or reader in this analogy) only sees the float as it goes by in front of him or her. Before the parade begins, the parade planner must anticipate how all of the pieces must fit together to give a nice show. They all have to fit well, and taking the perspective of being able to see the entire manuscript at once will help to plan for the best overall delivery. The author and the reader are different, and the author must take advantage of his or her perspective in order to give the reader the best possible outcome. Then, the parade can simply pass by much to the reader's enjoyment.

3.6 The Peer Review Process

While writing the actual manuscript is the hardest part, submission of the manuscript can be just as time-consuming even if less active. The journal for submission is selected. Then, the manuscript can be formatted, and the figures will be subsequently finalized based on how the journal prefers them. Always read the submission instructions.

However, the writing is not done. Each article needs a cover letter. This is the *MOST* succinct portion of the paper. Editors get tons of submissions. The cover letter must do two things: (1) tell the editor why this paper should be sent out for review and not rejected immediately by him or her; and (2) not waste his or her time. A good cover letter offers a cordial (if often meaningless) plea for consideration, a couple of solid statements as to why this paper is important (think leads from the Abstract and Conclusions), and a list of suggested references. While the last of these three are often handled as part of the electronic submission process in current times, giving the editor a heads-up is always appreciated. Sometimes, other information must be contained in the cover letter such as previous submission information, inclusion in a special issue, or other special instructions. These will help to guide the editor if such considerations are necessary. Always address the letter to the most senior member of the scientific staff like the Editor or Editor-in-Chief. The one caveat is if you know to which managing editor the paper will go, addressing the letter to him or her is acceptable. Find below a sample cover letter:

January 1, 2015
Dear Prof. John A. Smith, Editor, *Chemical Science*

Please consider the attached manuscript entitled "Complete Science Communication" for inclusion in *Chemical Science*. This manuscript argues that scientific communication should utilize the time-tested tenets of journalism for scientific writing. The inclusion of such techniques is shown to increase reader uptake and citations. This manuscript is being submitted as part of a special issue on the intersection of chemistry and the social sciences.

Potential reviewers include Dr. Neil DeGrasse Tyson of the Hayden Planetarium in New York City, Prof. Mark A. Littman of the University of Tennessee, and Prof. Richard Henderson from Corpus Christi College of the University of Cambridge.

Sincerely,
Ryan C. Fortenberry, PhD

There are two thoughts on suggested reviewers. An editor once told this author that if you suggest someone who is not well-known or not personally known to the editor, that individual will likely be invited to review the paper if no obvious conflict of interest can be established. Second, suggesting famous scientists within the field (such as those above) will likely be dismissed out of hand unless the submission is for one of the elite journals or the manuscript is immediately sellable as groundbreaking, warranting the expert opinion of such high-demand individuals.

Once the article is submitted (almost certainly electronically), then, the game is to wait. While passive, such time can be mentally taxing. The best advice is to distract oneself usually with the next project. Go ahead and start working on some other research whether it is directly related to the submitted work or not. This author once submitted two papers roughly simultaneously that were both not selected for publication for one reason or another. However, the resulting combination of the two items garnered a nice single manuscript that was readily accepted. The desire not to wait around led to the second paper and ultimately created a better combined manuscript in the end.

In this waiting game, a copyright release will likely be signed. While the author retains the intellectual property, the actual text may be owned by the journal (again, read the submission instructions for each journal). While the author can use the submitted manuscript (assuming acceptance) for his or her own purposes such as theses or presentations, the author cannot give away the article wholesale. Such is truly a small price to pay in order for the research to be disseminated to, and hopefully read by, the larger scientific community. There is more on this in the last section.

The editor will read through the important parts of the manuscript once it arrives in his or her inbox. Again, this is where the journalistic style comes in handy, especially in the Abstract. Editors are busy. If the author gives the editor what he or she needs immediately to judge the paper, then the readers will benefit from this readability, as well. If deemed potentially acceptable, the paper is sent out to typically one or two subject experts for peer review. They are given a time frame of two to four weeks to read the paper and offer feedback. From these comments, the editor may decide to reject the paper, invite the author to revise the manuscript, or recommend that it be sent to a "more appropriate" journal.

If the paper is in need of revision, the author should construct a cover letter that reproduces the reviewer's comments in full and addresses them pedantically and completely while describing how and where in the manuscript these changes have been made. Such changes should also be highlighted in an updated "marked-up manuscript for review." If the author wishes to argue against the reviewer, a full case should be made. However, simply doing what the reviewer suggests is always best. If that cannot be done in full, a compromise between the previous version of the paper and the reviewer's comments is rarely a bad road to take. From this author's experience, just do what the reviewer asks unless it really would downgrade the research being described. The revision cover letter should be addressed to the managing editor at this point. Their name will appear in any correspondence, and the author is usually notified of assignment of a managing editor shortly after initial submission. Resubmission then follows largely like the first submission.

Review rarely requires but can take several rounds, especially if the author and reviewer are in disagreement. Typically, the editor will side with the reviewer. If two reviewers examine a manuscript and give differing opinions, a third or adjudicative reviewer will be invited. This reviewer's opinions will really determine if the paper is accepted or not. Typically, though, a single vote for rejection by one reviewer is enough to kill a paper.

Rejection is not the end of the line for a manuscript. Submitting to a different journal is a viable option. If such a submission is to a journal within the same publisher's list, typically the previous review follows the paper. If it goes to a different publisher, the author is asked to volunteer whether the paper has been considered elsewhere. A candid statement regarding this is usually welcome, especially if accompanied by a concise statement from the author as to why the present journal is a better fit, which is always welcomed by the editor.

As an anecdote, this author was once asked to review a paper by a particular journal. The review's conclusion was to reject the paper. A few weeks later, this author was invited to review *the exact same paper*.

By "exact same," "verbatim" is the word including typos and incorrect logic conserved. In the second review, this author made it clear what had happened, and pasted in the first review. The paper was subsequently and hotly rejected by a very miffed editor. Hence, even if a paper is rejected by one journal, the manuscript should be updated since most reviewers will still offer constructive criticism. Once, the manuscript is ready for submission elsewhere, a truthful statement to this effect is appreciated. It will play to human nature, typically giving the editor a more favorable assessment of the document if the effort for improvement has been made.

Once, the manuscript is accepted, the author will be notified. The paper can then be listed as "in press" on the curriculum vitae. Often, if so selected by the author upon initial submission, the paper will be immediately posted in its unedited, submitted form online and will be given a digital object identifier (DOI). The paper can now be cited for posterity (and tenure packages). At some point thereafter, with the length of time depending upon the journal, the author will receive proofs of the paper for how it will appear in the "printed" form. In reality, this will simply reflect how the portable document file (pdf) format will look for those who download the article later. This is the last chance to make any corrections to the manuscript. If extensive changes are needed at this point, the publication will be slowed since the document will have to be type-set by the journal's staff or even sent out for peer review once more. Some journals publish the final form nearly immediately and some take many months. It just depends upon the journal.

3.7   Emerging Trends in Publishing and Writing Technically Without Peer Review

One of the biggest trends in modern scientific publishing is open access. However, it has opened a firestorm within the publishing community. Research funded by most government (including both that of the United States and United Kingdom) agencies requires that the results be freely available to the public. This is counter to the journal publisher's needs to at least break even if not make a profit. Hence, open access has emerged. The author, not the reader or the reader's institution, pays to publish. Granted page charges have existed for many reputable journals from the very beginning, but in such cases both the author and the reader had to pay. Open access puts all of the cost on the author at a much higher price.

Unscrupulous publishers have arisen where the article processing charge (APC) may start out low or not be listed at all. Then, once the paper is accepted and the copyright signed over, the publisher may raise the APC and hold the intellectual property hostage. These publishers should be avoided at all costs. The easiest way to navigate whether a publisher is reputable or not is to see if they publish other non-open-access or more traditional journals. If they do, they are likely a credible organiztion that will not hold your work hostage for sordid gain. If they only publish open access, ask a librarian or a trusted colleague. *RSC Advances*, for instance, is an open-access journal operated by the Royal Society of Chemistry. While they charge an APC to publish in such an open-access journal, their business was built in the traditional way with most journals requiring subscriptions. Subscription-based journals are free for the author to publish. Hence, if a publisher has both, they are most likely credible and will give the author options about open access.

One way that the need for open-access can be circumvented is through online repositories. The most famous is the arXiv operated out of Cornell University. Authors may post versions of their manuscripts on this server for others to download for free. This violates some publishers' copyright agreements. It does not for others if the uploaded version is not the final version, *e.g.* not a photocopy of the proofs and similar. Some publishers, like the Royal Astronomical Society, encourage authors to submit to arXiv because they view it as free advertising, and they are a non-profit organization simply trying to make enough money to cover costs. Still, an author should carefully read the copyright agreement before posting papers to such repositories. These should not be confused with websites like http://www.researchgate.com, which charge memberships or cap the number of free downloads. Uploading papers to such websites that turn profits off of published materials are almost always in violation of copyright.

In any case, standard repositories like arXiv have revolutionized astrophysics and planetary science. Chemistry has been somewhat late to the game. While many interdisciplinary papers involving chemistry, especially chemical physics can be found in such formats, dedicated chemistry repositories are relatively new. The American Chemical Society, along with the German Chemical Society and the RSC, have begun their own repository, https://chemrxiv.org/, where the specific copyrights for their journals are not violated.

One thing about these free, crowd-sourced repositories is that they are not peer-reviewed. While most scientists have these manuscripts attached to peer-reviewed publications and use things like arXiv in order to generate traffic to the paper or for early alerts regarding exciting discoveries, anyone can post nearly anything. Hence, manuscripts that cannot be published in traditional fashions are often uploaded, and crazy ideas can be posted. Consequently, such repositories should be viewed in the same way as encyclopedias in journalism. They are secondary sources and great places to get ideas, but the actual citations and publication should be done in more traditional ways.

Finally, social media should be utilized to promote research publications, and this is discussed later in the publication relations chapter. In brief, though, a departmental or research group Twitter account or Facebook page are great ways of connecting with fellow scientists, supporters, or prospective students. Every article published should be linked either *via* such an account or from the scientist's own personal social media handle. In the age of Google where links determine priority in a search, having more links from self-promotion is never a bad idea. Plus, this increases serendipity and the chances of something interesting being seen by someone who may not otherwise know about this research. Again, this is discussed at length in Chapter 6.

3.8   Assignments

3.8.1   Poetry

Have the students choose both a poem and a peer-reviewed, scientific journal article. Have them read the poem, then the article, and then the poem again. This can be discussed in class or given as a written assignment. Hip-hop, sung, or lyrical verse should not be used in place of poetry written as poetry. In the case of the written assignment, have the students do the following.

Assignment for Students: Write a two-page response in proper Royal Society of Chemistry style to this exercise. Give the name and author of the chosen poem as well as the citation for the peer-reviewed article. Respond to the following thoughts: (1) Why was the given poem and research article chosen? (2) How did it feel when going from the poem to the article? (3) How did it feel going from the article to the poem? (4) How can a scientific piece incorporate the styles of poetry?

For the instructor: Typically, poetry is easy to read. It evokes emotion, is free to interpretation, and contains structure. Hence, the shift from poetry to technical writing will illicit disonances. However, the shift back to poetry should feel like a relief. The reading should go from, "Here is the information; take it" in the technical article to "Choose from this what you will" in the poem. Encourage the students to employ the word choice, sentence structure, and more open interpretation when writing the Introduction and Results section. Also, encourage them to read poetry (or other stylized language) during the semester.

Suggested Poems:

- Phenomenal Woman by Maya Angelou
- I, Too by Langston Hughes
- Musée des Beaux Arts by W. H. Auden
- Crossing the Bar by Alfred, Lord Tennyson

3.8.2 Reading Abstracts

Assignment for Students: Find two abstracts published in either *Chemical Science* or *Nature*. Choose one that you feel is well written and one that you feel is not well written. Write a two-page discussion (in total) about the strengths of the one you thought was good and the weaknesses of the one you thought was poor. Also offer comment on the good items in the bad and vice versa for each abstract. Draw, most notably, from how a lead was or was not employed, as well as whether or not clear descriptions of the Methods, Results, and Conclusions are also included. Comment on the structure, readability, and length of each abstract. This can also be done as a class discussion.

3.8.3 Writing the Abstract

While the Abstract is typically the last thing to write in a paper, it makes a nice entrance into technical writing. In truth, the main goal of technical writing within science is to write research articles for scholarly journal publication or proposals for time on specialized equipment or money to do research. Even though technical writing is different from journalistic composition, utilizing some points of the latter enhance the former.

Assignment for Students: Utilizing your current or past research, you will write a 150–250 word abstract for submission of your research to a symposium for a poster or oral presentation. Even though the length may not be significant, the density of information is. As such, you must be exceptionally clear, concise, and correct.

Grade this assignment as follows:

- Choice of topic (10%)
- Length (10%)

- Proper motivation and attractiveness (10%)
- Density of information (20%)
- Clear, concise, correct writing (50%)

3.8.4 Writing the Methods and Results Sections

Assignment for Students: In technical writing, the most important part besides the lead is the detail that allow your research to be reproduced. The Methods are typically also the easiest for one to write. Here, you will write 250–1000 words properly describing the technique(s) utilized to perform your research experiment. Again, utilize your current research project or a previous one you have done. Only do a previous one if you are not currently engaged in research. Then, you will construct data tables containing the results of your research. Finally, you will write an appropriate length section for the results (what the data are) and discussion (what the data mean) of this research. You need to synthesize your ideas and give the non-expert but still specialist in your field reading this enough information to understand what is going on in order to build upon this work for his or her follow-up research.

Grade this assignment as follows:

- Development and Completeness in the Experimental/Computation Procedure/Setup/Details (10%)
- Length of Experimental/Computation Procedure/Setup/Details (10%)
- Proper construction of data tables (10%)
- A necessary but not too lengthy description of the data (10%)
- Synthesis and "so-what" discussion (20%)
- Clear, concise, correct writing (40%)

3.8.5 Writing the Introduction While Building the Discussion

Assignment for the Students: In techincal writing, you must show the motivation of the work that you have done. You are to write an Introduction section for the research that you wrote about in your Methods & Results sections in the previous assignment. You will append the updated Methods section to the Introduction here, but most of the grade will be for the Introduction.

Here you will need to construct a narrative that motivates your research while explaining all of the things done in the past that have led to your ability to conduct this experiment. You should basically think of this as a 500 word news piece about the foundations of this research project. You will need to use proper Royal Society of Chemistry style for citations and have a developed set of references that sufficiently develops the research at hand. Again, you need to synthesize your ideas and give the non-expert but still specialist in your field reading this enough information to understand what is going on in order to build upon this work for his or her follow-up research.

Grade this assignment as follows:

- Proper use of citation style (10%)
- Proper utilization of the background literature (10%)
- Proper bridging of the past with the present while only setting up the present in the Results (10%)
- Updates to the Methods and Results (20%)
- Clear, concise, correct writing (40%)

3.8.6 Full Paper

Assignment for Students (Major): Write a scholarly paper for submission to a chemistry/physics journal such as *Chemical Science*, *Physical Chemistry Chemical Physics*, *Journal of Materials B*, *etc.* You must include an Abstract, Introduction, Experimental Details, Results & Discussion, Conclusion, Acknowledgements, references, tables, and figures. The Introduction says why anyone should care and gives a description of what has been done. The Conclusion ties everything together with a pretty bow. The abstract is what most people who look at your paper will read; it needs to hook people in. You also need a cover page for submission to the "journal." Again, the experimental setup must be complete and reproducible, while the Results talk the tables/figures and the interwoven Discussion explains the data.

Grade this assignment as follows:

- Adequate motivation for the work in the Introduction (10%)
- Adequate literature discussion in the Introduction (10%)
- Development and Completeness in the Experimental/Computation Procedure/Setup/Details (10%)
- Proper construction of data tables and/or figures (10%)
- A necessary but not too lengthy description of the data (10%)
- Synthesis and "so-what" Discussion (10%)
- Fitting Conclusions (10%)
- An engaging Abstract (10%)
- Clear, concise, correct writing (20%)

4 Speaking (not) Like a Scientist

4.1 Introduction

Giving an oral presentation is not like writing a journalistic piece. This is a fireside chat and should, hence, actually tell a story. Humans throughout our existence have built relationships around telling stories and sharing tales, in varying degrees of truth. These interactions build communities and also educate one another through vicarious shared experiences. This is why culture and communication are so inextricably linked. The written word is more like a personal gleaning of information from the surrounding areas, but public presentations are communal, campfire tales. The former is informative, but the latter is relational. The speaker's job in an oral, scientific presentation is to build a relationship, however brief, with his or her audience. Taking cues from man of old is the best way to communicate modern scientific knowledge.

In fact, Walter Fisher, a modern communication theorist, states that the best communication is not built on logic but on a good story. His "narrative paradigm" states that the most convincing argument is an engaging story that contains both fidelity (how well the story rings true) and coherence (how well the story fits together). This type of model fits best within the realm of communication for science within the oral presentation. Hence, the most convincing scientific argument, especially if it is done in a person-to-person or person-to-audience format, is an engaging storytelling experience.

Complete Science Communication: A Guide to Connecting with Scientists, Journalists and the Public
By Ryan C. Fortenberry
© Ryan C. Fortenberry 2019
Published by the Royal Society of Chemistry, www.rsc.org

Our modern presentations have their own built-in campfires: the flickering screen. The job of the scientific presenter is to utilize this visual aid in much the same way that the campfire has been used from antiquity. This flickering light keeps the listener engaged but not overwhelmed. It allows the listener to lose focus on the presenter for a time while not losing focus on the story being told. However, this natural human tendency mixed with improper use of presentation slides can lead to "death by PowerPoint" where the campfire becomes everything and the speaker is a useless voice over the crackle of the fire. Like with anything, balance between the speaker and visual aid is key.

As a final note of introduction, no one gets upset about a presentation that ends early. Everyone gets upset about one that goes long. There is an old saying about speaking (particularly in church): "Stand up to be seen; speak up to be heard; and shut up to be appreciated." Once the allotted time is surpassed, the members of the audience truly stop paying attention. They start thinking about how they are late for the next class, how little food is left in their rumbling stomach, or why this jerk behind the lectern will not shut up since no one really cares any more. In truth and to reiterate, no one cares at the point of going on too long. The remaining presentation simply becomes a waste of time for everyone. Truly, "shut up to be appreciated" no matter how important you think you are.

4.2   The Art of Presenting

The most important single thing in any presentation is the presenter. The slides are second. The stage is second. The environment is second. The most important aspect is the man or woman pontificating. Nothing should distract from the presenter, even, maybe especially, the visual aid. Watch a TED talk, arguably the best brand of scientific presentation ever. The focus is undeniably on the presenter, and they have very few, very stark slides. The folks at TED figured this out years ago. The rest of the scientific community must catch up.

The best presenters keep their audiences in tune with them and not with anything else. The speaker is the central focus. This affects slides, sure, but it also influences dress, general appearance, demeanor, word choice, structure of the narrative, *etc*. The fewer distractions, the better. This is why theater is performed in a darkened room with bright lights on stage so that the action taking place on the stage (or being hinted occurring off-stage) is the primary focus. Much of the play takes place not on stage but in the audience members' minds and is guided by the narrative being told. The actors merely provide cues to the audience to create the story in each member's own mind. Focus on the performance allows the brain to wander into new places and experience a richer depth of the performance than what the action on stage provides. For instance, in the play *Harvey* by Mary Chase later turned into a movie of the same name starring actor Jimmy Stewart as the protagonist Elwood P. Dowd, the titular character never actually appears on stage. However, the entire story is constructed so that Harvey's presence is real to the audience without them ever seeing him. The focus on the stage action drives the audience to imagine Harvey right along with Dowd. This clear focus on the stage characters hints at, but never fully reveals, Harvey. The best presentations lock the audience members' gazes but free their minds.

The speaker/presenter is the master of the stage even if that is the only place he or she masters. Stories in Hollywood say that off-stage Johnny Carson, the American late-night television master, was quiet, reserved and mostly enjoyed talking with Ed McMahon at parties. On-stage, he came alive and held his audience in the proverbial palm of his hand. While Carson naturally had a gift, his command of the stage and love of the audience on-stage was balanced by his personal shyness and desire for privacy and personal space off-stage. Hence, even the most stereotypical, introverted scientists can embrace the performance of the presentation.

The key in presenting is to remember that no one in the audience knows more about the topic than the presenter, and the narrative is the speaker's story to tell. If the one giving the talk keeps this personal omniscience in mind, he or she should find a confident peace. Granted a few audience members may have expertise that lets them ask difficult questions or may be well-versed in a particular aspect, but none of them were in the lab doing the work getting the numbers and findings. That is the speaker's trump card.

Carson knew the punchline before the audience. Some could guess what it might be, but only he *knew* the answer. This gave him the edge, and he could utilize this to make the audience respond as he wished. The scientific presenter knows the story he or she is going to tell. Having the confidence, and to a small extent swagger associated with it, of withholding something the the audience wants and being in command of when they get told should make the presenter the most interesting person or thing in the room. This is the core to the art of presenting: the speaker is important because he or she knows a little secret but will only divulge when he or she wishes. The best audience is that which is kept enthralled waiting on the delivery for as long as enjoyably possible, but not longer.

The best way to keep the audience enthralled is never to give them a chance to feel as though the speaker is disengaged. In other words, *ALWAYS* face the audience and *NEVER* stare at the slides. A quick glance at a slide is necessary to maintain correspondence between what is said and what is on the screen. However, the most important thing on stage is the presenter. Maintaining dynamic eye contact is key. The speaker should look the audience members in their eyes. The gaze should bounce around the room and settle on a new person each time. If possible, the speaker should get out from behind the lectern and speak from different positions on the stage. Getting out into the audience, especially with the aid of a remote slide advancing device, can be tricky, but if done correctly, the audience engages with the speaker more readily since someone close is more human than someone far away. The physical actions of the speaker should always be done with a mindset of making the audience feel on an increasingly personal level with the one giving the presentation. This will encourage the asking of questions which leads to a discourse and a more memorable experience for the questioner. The speaker should smile and enjoy him or herself. This is his or her show as the speaker, and it should be obvious to the audience.

Finally, the speaker should never say "um" or other verbal pauses like "here," "so," "you know," or "what?" While these are often unavoidable, they should be minimized. These serve to let the audience know that the speaker is thinking and does not want to be interrupted. Sometimes in interpersonal communication such uses are acceptable, but they are not in public speaking. The speaker's phrases and sentences should be clear and to the point. Plus, if a moment is required for the speaker to think, strangely enough, a brief silence is actually more acceptable than a stammering of "um" and "uh."

4.3   Structure of the Presentation

The traditional storytelling plot diagram, given in Figure 4.1, of Setting, Rising Action, Climax, Falling Action, and Resolution is finally proper in an oral presentation. Again, this is a campfire story. In some ways this follows the typical Introduction, Methods, Results, and Conclusions of the written scientific paper. However, these traditional pieces can easily turn into crutches for the presenter. The best speakers drive the presentation–are not driven by them–and utilize Setting, Rising Action, Climax, Falling Action, and Resolution within

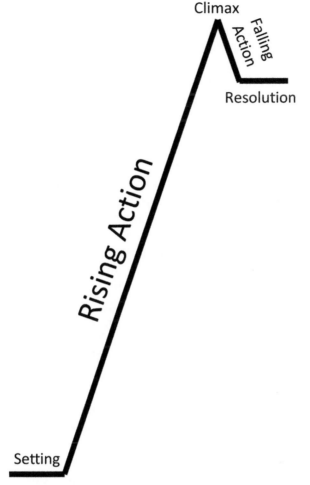

Figure 4.1 The standard plot diagram with the lengths of the lines for each portion roughly equating to how much time should be spent on each portion of the narrative.

the context of a scientific discussion to create a compelling narrative just like Fisher advocates for persuasion.

There are clear characters (which are not the scientists doing the research). These characters are the items that are the subject of the study, and their stories, whether for good or ill, are developed throughout each stage of the story. These characters drive the story and the elements therein. For instance, if the research is on toxins in the water supply, the story could be given a feel of a spy novel where the infiltrators are present, but the "good guys" are having to utilize all possible tools available to find them. This may seem silly, but having this perspective will make the story told more entertaining.

4.3.1 Setting

The introductory material provides the Setting succinctly. Take the prologue to Shakespeare's *Romeo and Juliet*:

> Two households, both alike in dignity,
> In fair Verona, where we lay our scene,
> From ancient grudge break to new mutiny,
> Where civil blood makes civil hands unclean.

The entire situation is laid out clearly. The audience is provided the perfect setting for a tale of love and loss. However, this is done concisely; note the length of the line for "Setting" in Figure 4.1. The setting does not list which ancestor murdered which other ancestor. It does not cite the source of how the families know of their ancient feud. It elegantly paints a picture that is necessary for the plot but is not overly verbose. The story is not about what happened to the Capulets and Montagues in the past. The story is about Romeo and Juliet (hence the title) now. Verbosity is left for textbooks and encyclopedias where such detail can be looked up later and is not proper for the spoken word. As a result, the Introduction of an oral presentation provides the Setting for the story and mentions what is necessary to open the story, but this does not become the story. It does not dither on the details, but "just gets on with it," to use a common English idiom. Like the tragedy of events surrounding Juliet and her Romeo, the actual story in the scientific presentation is what the characters do and the adventure they have.

There are two ways to take the discussion of the Methods. These could be part of the Setting and, hence, Introduction. Or they could be woven into the story itself, arising in the natural progression of the story. This author uses a mixture of the two where he lays out his overarching procedure in the beginning and gets it out of the way as part of the Setting. Then, when specific methodological details arise as part of the story, they are produced in an *ad hoc* basis. This goes along with telling the story on a need-to-know basis. Setting only works when it can be retained in full by the audience.

4.3.2 Rising Action and Climax

The narrative of the presentation should be built such that a Climax of most vital information is provided. The struggles, pathways, dead ends, minor successes, *etc.* all make up the story in the Rising Action;

note the length of this portion of the line in Figure 4.1. The easiest mental construct for setting up a good story even if the subject is about scientific research is in telling a friend about the journey had while getting lost on the way to her house. This path is taken; that path is taken; an illegal U-turn had to be employed. There was a policeman wearing a dachshund on his head who helped to point out the proper way forward, and, then, finally, the house came into view with a great sigh of relief. Again, the speaker is telling a story around a campfire and all the elements of Figure 4.1 are there. The Setting is made, the Action builds to a Climax, and then the most important Result of arrival is finally achieved.

The Climax is actually not an event but is simply the transition from an increase in intensity of the story to a decrease in intensity. To reiterate, the Climax is not an actual event but simply the shift in the type of events taking place. The Climax is merely an instantaneous transition just like the brief moment when an upward thrown ball stops before descending back toward the Earth. After this turning point Climax, the Action falls with some odds and ends of other results, and then the presentation ends at a satisfying Resolution.

4.3.3 Resolution

The Resolution is a satisfying ending and *NOT* a summary of the key points like that in the Conclusions of a technical, peer-reviewed paper. This author used to have a slide of bullet points summarizing the key findings. He quickly realized that this slide is mostly just a point at which the audience can relax and tune him out. The best conclusions are left in the audience member's minds. Beowulf does not stand on a kopje to recite his great deeds at the end of the epic bearing his name, nor does a eulogist expound upon the hero's life's worth once he is killed. Instead, the hero simply dies and is burned at sea per tradition. The reader knows the story already and is satisfied with the Resolution of a great life. The Resolution should be a natural ending to the story told and not a redundant summary.

Then, a brief mention of acknowledgements should always be included. Items to incorporate are the people who did the work in addition to the speaker such as students, collaborators, or mentors; the funding agencies or individuals; and any further, non-scientific notes that need to be made. Some speakers like to put these at the very beginning for the Introduction and Setting so that such people are not missed. Really, these should belong at the end provided that the allotted time is not gone over.

4.4   The Visual Aid: The Campfire

A visual aid is not a visual crutch. The aid exists for the audience and not for the presenter. However, bullet points, lengthy paragraphs, and walls of text often insulate the presenter from actually having to remember the story being told and also prevent him or her from having to do anything creative. Such a sophomoric style has the speaker facing the slides, reading the material. This type of poor presentation style is the visual crutch and should be avoided at all costs. It has permeated scientific presentations for decades often being the butt of many jokes. While this is changing, the time has come for such poor presentation skills to die.

The most important thing in a presentation is, again, the speaker. The visual aid augments his or her words. It does not repeat them. The brain processes the written word and the spoken word in the same place. Hence, language, regardless of its medium, can only come in through one channel. As a result, the audience will either read the words or listen to the words. They will not do both. Visually inclined audience members will tune out the speaker to read the words, making the presentation a waste of time for both. Members of the audience who prefer to listen will only hear the words, making the slides useless as a visual aid. The heuristic is that words are spoken, not written, in an oral presentation.

The visual aid provides for the speaker what his or her spoken words cannot. Pictures and animations are the most important aspects of what to include in a presentation. Pictures are truly worth a thousand words. Animations of those pictures are worth a few hundred more. The most straightforward and base-human means of illustrating this point is in a human face. Words fail in simply trying to describe another human face. "The eyes are blue but almond-shaped. The nose peaks downward." This could go on and on. However, simply putting up an image of a face removes in less than one second the need for unnecessary and inadequate description. Images in the visual aid are like putting up a picture of your cute niece instead of trying to describe how cute she really is. Hence, use this heuristic for knowing when to speak and when to simply display an image: if it takes more than ten words to describe an item, show a picture.

Data tables are also items that are more conducive to visual depiction than oral explanation. The key in utilizing these often necessary but overwhelming splatterings of numbers is that walls of text are assimilated in the human brain into the larger image, the proverbial trees for the forest. The speaker must highlight the key findings within large associations of information. Highlighting bars or glowing text can accomplish this. Really, any useful figures, images, cartoons, and charts go well on slides. Text does not. Consequently, if lots of time and spoken words are spent to explain an idea, visual images should be provided for the audience.

If the point is not abundantly clear enough, slides should have no text on them. Any words that would otherwise be needed should come from the speaker. In some instances, a caption for an image can help, but this should not be overused. In truth, this author avoids such. Additionally, some slides work well with slide titles like "Introduction" or "Results." However, many times they are unnecessary and are often ignored by the audience anyway. This author uses a mixture of including and excluding slide titles depending upon the progression of the presentation's narrative. Typically, titles are only used at the beginning or to distinguish items in the various stages of presentation. The animation of words can be useful since the audience member's brain has already taken the word and translated it from the symbol into the meaning. Animating the word in some way is similar to, although not quite as good as, animating a corresponding image.

Animations are also an excellent way of keeping the reader involved and allowing the campfire to flicker in novel ways. In the very least, no picture, figure, or other slide item should be present on the slide unless it is the current topic of discussion (or remains on the slide after having been immediately and previously discussed). Putting all of the visual information on the slide at once and then going through it makes the listener bored. He or she has already processed through the information and tuned out the speaker altogether (at best) or has become completely confused why any of this stuff is on the screen (at worst). Each item on a given slide or each slide should be doled out as the speaker is discussing each topic. The words spoken by the presenter are still only in the present, not the past or the future, tense. The slides should be in the present, as well.

Animation motions can become overwhelming quite quickly. The best animation is really a simple fade of about half a second. This is simple and to the point. The image appears in a soft manner that does not surprise the audience and allows for a newly established thought. Sometimes large animations are quite useful, especially if the object is moving or if added emphasis is needed. For instance, the Stratospheric Observatory for Infrared Astronomy is a former jet-airliner re-purposed by NASA and the DLR with a hole cut in the side for an infrared space telescope mounted inside the airframe. When introducing this telescope for applications of astrochemistry, the photo of the plane flies onto the screen as it would through the sky making a natural motion that is not highly distracting. Increasing scale or putting colored highlights behind an image or data point are also great ways to emphasize important aspects of a slide. Additionally, quickly fading different chemicals onto a slide and moving them all into a flask is a great way to show the combination of materials for a solution to be analyzed or a reaction to take place. The key is utilizing the motions of the images in a natural way and not simply animating for the sake of animating, which becomes distracting.

In making images to put on the slides, the best thing to do is try and isolate the desired part of the image as much as possible. Hence, the background removal tool on most visual aid software (such as Microsoft PowerPoint or Apple Keynote) is highly useful. This reduces the visual noise. It may take some practice and is a bit of an art as portions of the image can be lost that are needed or retained that are unnecessary. However, utilizing only the pertinent portions of the image is essential. In some cases, removing the background is too difficult or, honestly, weird. For instance, a fluorescing vial could look out of place by itself without the dark background of the rest of the lab. Additionally, photographs of people can be strange when cut out of the image. In any case, standard, square cropping of the image is essential to remove as many extraneous portions of the image as possible. Regardless, images/photographs with some background should utilize a nice, thin, black outline around the image. This helps to distinguish the photograph from the rest of the slide.

Additionally, shadows can be added to images for further emphasis and three-dimensional rendering. This makes it easier for the audience to perceive that which is important. Something with a long shadow is likely closer to the audience member making it more important than something farther away. These make a depth to the slide that is natural for the audience and really makes the images pop. Even just adding the default setting for shadows will help with this effect.

One thing that absolutely needs words are the references. These should be as short and concise as possible at the bottom-left of the slide. Sometimes just the journal, volume, and page number are enough. However, this author likes to include the authors on the papers (as long as they are not too gratuitous) in order to acknowledge the work done, especially if it is work done by his students.

Honestly, the best slides have white backgrounds and (when absolutely necessary) black text. Any other combination is, frankly, not always guaranteed to show up and will at times look really terrible.

Quantum Chemistry and Spectroscopy: A Match Made in the Heavens

Ryan C. Fortenberry, Ph.D.

Asst. Prof. of Physical Chemistry

University of Mississippi

State University, Month ##, YEAR

Figure 4.2   A sample title slide.

Keep the slide scheme simple. The images will have color, and the speaker's story will be engaging enough.

4.5   Sample Presentation

The sample title slide in Figure 4.2 has lots of words. This will mostly be put on the screen when the audience is filtering in or getting settled whether in a seminar or a symposium talk. Largely, the title slide is a place holder, but it (naturally) gives the title of the talk, the presenter's name, and his or her affiliations. The rest is really style. The place and date of the presentation can be put on there, as well.

The header at the top remains constant for this slide and gives what is hopefully a novel but not too distracting standard item for use on all slides. The logo for the author's affiliation is in the upper-left and images associated with his research keep the title bar going.

At this point the presenter will be introduced either by the inviting faculty member if a seminar or the session chair if at a symposium. The presenter can say some words of welcome and such in order to be a gracious speaker. Often, this author attempts to establish some common ground, soften the audience, and/or give acknowledgement for the invitation/ability to speak in the venue. When presenting a

Astrochemistry

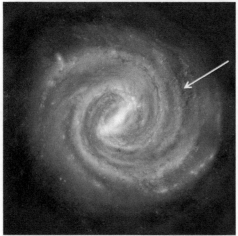

You are here.

Figure 4.3 Setting slide 1 stage 1.

single seminar, he will show a photograph or give an anecdote either about the one who invited him or about the place where the seminar is being held. This is a matter of style, but it shows the audience that the speaker is not just some disengaged scientist who came in to impress everyone with his or her work. This can be overdone, but at least a word of thanks for the opportunity to speak is warranted at this point.

This sample presentation will incorporate major portions of a talk given on the quantum chemically derived rovibrational properties of proton-bound complexes for application to their detection in space. While the topic is somewhat esoteric, the general points and examples should still be quite useful for the reader.

4.5.1 Setting

This trio of slides in Figures 4.3–4.5 are actually stages of one slide. First, immediately after the title slide, an artist's rendition of the Milky Way galaxy is shown in Figure 4.3. This establishes the scope and setting of the story promptly followed by emphasis on the arrow and text that appears on the next mouse click. Yes, this is text, but this text makes a point about the minuscule nature of humanity. This

Figure 4.4   Setting slide 1 stage 2.

text is also not read out, but left for the audience to see. Then, in Figure 4.4, the Milky Way is further expanded such that it fills nearly the entire slide (and, hence, screen) in order to further emphasize the vastness of the medium for astrochemical research. The last stage in Figure 4.5 has more words. However, this is a quote and would be something stated anyway. The fact that this has been printed in a commonly read, popular science magazine (*Scientific American* in this case) adds credence for the presenter as an expert in the field. This quote is recited to the audience without the speaker even looking at the screen. This recitation is often done from the middle or even back of the audience for emphasis. Again, this is a matter of style, but the point is that the speaker's words and actions married to the visuals from the campfire of presentation slides build a firm foundation for the setting of the oral talk.

The Setting is further established by Figures 4.6–4.8 which are, again, differently built stages of the same slide. Figure 4.6 is introduced as a spectrum taken from the interstellar medium. Overlaid on this is the visible spectrum in order to show the audience where the given peaks occur. A perspective shadow is given to the spectrum

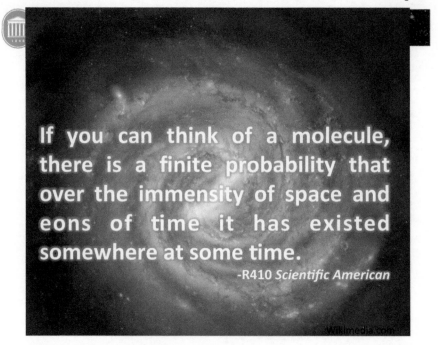

Figure 4.5   Setting slide 1 Stage 3.

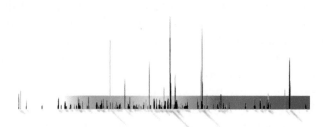

Figure 4.6   Setting slide 2 stage 1.

Figure 4.7 Setting slide 2 stage 2.

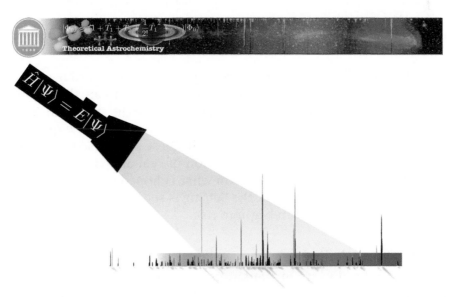

Figure 4.8 Setting slide 2 stage 3.

in order to make it have a tangible feel. This spectrum, called the diffuse interstellar bands or DIBs, is described by the speaker and all pertinent details are given.

The next build stage in Figure 4.7 brings in a clipart (made by this author in order to avoid copyright issues) in the shape of a flashlight.

Computing the Anharmonic Potential

$$H = \frac{1}{2}\sum_{\alpha\beta}(J_\alpha - \pi_\alpha)\mu_{\alpha\beta}(J_\beta - \pi_\beta) - \frac{1}{2}\sum_{k}\frac{\partial^2}{\partial Q_k^2} - \frac{1}{8}\sum_{\alpha}\mu_{\alpha\alpha} + V(\mathbf{Q})$$

$$V = \frac{1}{2}\sum_{ij}F_{ij}\Delta_i\Delta_j + \frac{1}{6}\sum_{ijk}F_{ikj}\Delta_i\Delta_j\Delta_k + \frac{1}{24}\sum_{ijkl}F_{ikjl}\Delta_i\Delta_j\Delta_k\Delta_l$$

X. Huang and T. J. Lee, *J. Chem. Phys.* **129**, 044312 (2008).
R. C. Fortenberry, X. Huang, J. S. Francisco, T. D. Crawford, and T. J. Lee, *J. Chem. Phys.* **135**, 134301 (2011).

Figure 4.9   Rising action slide 1.

This flashlight has the Schrödinger equation on it in order to show that there is a tool in quantum mechanics which can then "shine light" (to use an idiom), as given in Figure 4.8, on the spectrum. These visuals combined with the speaker's words highlight how quantum chemistry will be utilized to help explain these spectroscopic mysteries of astrochemistry. This combination of oral and visual messages work in tandem to communicate in what is hoped the most effective means possible. At this point, the Setting is set (and really in only two slides), and the Action can begin to rise.

4.5.2   Rising Action

There is no bright line in transition between the Setting and Rising Action, at least as far as the oral tradition is concerned. Some might consider early stages of the Rising Action still to be part of the Setting, and other scholars may not. In practice and for us as science communicators, it truly does not matter. However, this author views the transition as: once the stage is set, anything that involves the present research is the beginning of the Rising Action. Again, the Rising Action will be a majority of the talk and story to be told, as per its line in Figure 4.1.

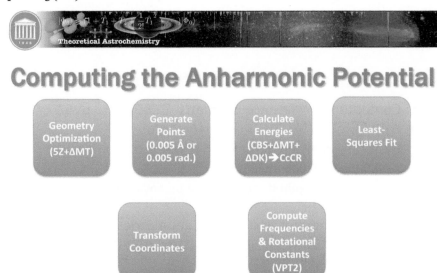

X. Huang and T. J. Lee, *J. Chem. Phys.* **129**, 044312 (2008).

R. C. Fortenberry, X. Huang, J. S. Francisco, T. D. Crawford, and T. J. Lee, *J. Chem. Phys.* **135**, 134301 (2011).

Figure 4.10   Rising action slide 2.

In this sample presentation, the Action rises initially with an explanation of the computational methods employed. The necessary mathematical equations are produced in Figure 4.9. These equations are big and intimidating. They are not intended to be understood by the audience if those listening to the presentation do not already know how to read them. A superficial explanation as to the mathematical expressions given in the left of the top equation in Figure 4.9 is stated that these are merely kinetic energy terms. The potential, "V," is defined in the lower equation. Really, these equations simply exist to show that some form of an equation must be fit in this research to go along with the Schrödinger equation depicted on the previous slide in the Setting stage. A detailed explanation would not give the audience members who are familiar with this equation (or a similar form of it) any useful information, and it would not give any audience members not familiar with it a real hope of understanding. Hence, such explanations are often unnecessary time sinks.

Specifically, this equation is introduced by utilizing the procedure on the next slide, Figure 4.10. This figure has words, but they exist to

Computing the Anharmonic Potential

- Compute the reference geometry: CCSD(T)/cc-pV5Z with CCSD(T)/MT for core-correlation.
- Generate displacements from reference geometry:
  - 0.005 Å or 0.005 radians.
  - Up to a 4 magnitude sum change on any of the coordinates.
- Compute energies at displaced geometries:
  - CCSD(T)/CBS (T,Q, & 5Z) + CCSD(T)/cc-pVTZ-DK + CCSD(T)/ MTcore to give the CcCR QFF.
- Least squares fit of energies to form the QFF & VPT2.

X. Huang and T. J. Lee, *J. Chem. Phys.* **129**, 044312 (2008).
R. C. Fortenberry, X. Huang, J. S. Francisco, T. D. Crawford, and T. J. Lee, *J. Chem. Phys.* **135**, 134301 (2011).

12

Figure 4.11   Rising action alternative slide 2.

provide tangible information for the audience about how the procedure is undertaken. An earlier version of this slide is given in Figure 4.11. However, the text is given in bullet points and is far from ascertainable. In the case of Figure 4.11, the presenter will essentially be reading the slide, and the audience will likely either tune him or her out or will not even look at the slide. Either case results in the discussion being pointless. Hence, Figure 4.10 is the better choice for displaying this information. Furthermore, each of the boxes in Figure 4.10 are faded in as the points are described. This highlights the necessary procedure. The audience will almost certainly not remember (and likely not care) what the procedure utilized is, but, again, this establishes credibility with the audience.

In using the analogy of classical literature, this type of beginning to the Rising Action is akin to a major character taking part in a short anecdote to get the story off the ground. In returning to *Romeo and Juliet*, Abram picks a fight (by biting his thumb) with Sampson and Gregory in order to establish to the audience the extent to which the insanity of the feud has taken. These characters are not part of the

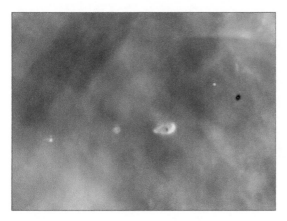

Figure 4.12   Rising action slide 3 stage 1.

rest of the play (at least not in name), but this is a transition from the setting into the present story.

The story continues to rise as more specificity is brought into the action at hand. Figures 4.12 and 4.13 provide more context for the story. These are photographs of protoplanetary disks where baby solar systems are created. These images are great ways to get the story going. Images of big picture applications, solutions, or outcomes for the research at hand work well in telling the audience where the story is headed. Again, images are truly worth a thousand words. This gives a depth to the story being told and a visual idea about the narrative. These protoplanetary disks could be described in thousands of words, without the audience truly understanding what is going on like they would with a photograph from NASA's Hubble Space Telescope.

The story is then brought back to the familiar with an image suggestive of the periodic table in Figure 4.14. A full periodic table could have been given, but there is no need. This shape and coloration should be familiar enough to any audience, even non-scientists. Additionally, the lack of details allows the viewer not to have to process it deeply, allowing for quicker uptake of the image. However, a certain portion of the periodic table is needed for discussion as shown in Figure 4.15.

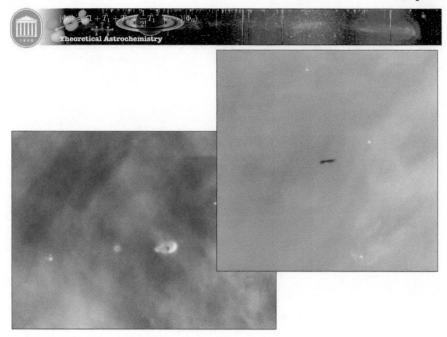

Figure 4.13   Rising action slide 3 stage 2.

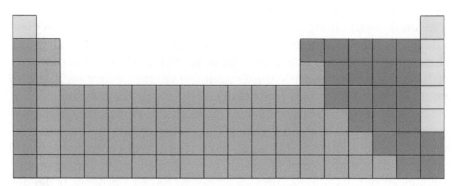

Figure 4.14   Rising action slide 4 stage 1.

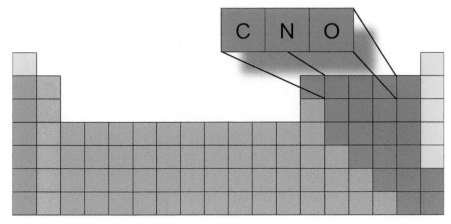

Figure 4.15 Rising action slide 4 stage 2.

Now, only a specific part of the table is required, and these items are brought out to be highlighted. The juxtaposition of Figure 4.15 to that from Figure 4.13 indicates that molecules containing these atoms are of interest for study in regions of space such as the previously highlighted protoplanetary disks.

This is brought full circle with Figure 4.16. The same image from Figure 4.12 is highlighted by being included as a full-screen depiction showing the baby solar systems. Then, the nitrogen molecule is faded onto this image (not shown in stages) because the search is for the nitrogen molecule. At this point the discussion becomes about the nitrogen molecule with no further need for slides or bullet points. This image frozen in time is enough for the presenter to describe that the typical astronomical tools of microwave and infrared spectroscopy cannot work for N_2 because it possesses no permanent dipole moment or even an inducible dipole moment. While electronic spectroscopy could be used and even claimed to detect this molecule in space, this is fraught with challenges. Hence, with over three-quarters of the atmosphere made up of this seemingly invisible material, how can this substance be observed in baby solar systems where primordial Earths may be forming?

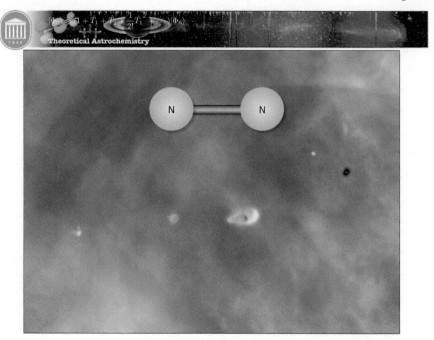

Figure 4.16   Rising action slide 5.

None of this discussion is shown in pictures since images of the Green Bank Telescope in West Virginia, USA or the upcoming James Webb Space Telescope would largely be distracting from the presenter. The story stays on the nitrogen molecule since the most important things to be discussed are coming from the speaker. He or she is asking the audience to become emotionally involved in the plight of the nitrogen molecule. The faux damsel in distress is employed to elicit this connection, done through traditional storytelling.

For example, in a classic scene from the 1975 film *The Great Waldo Pepper* starring Robert Redford as American, barnstorming and trick-flying biplane pilot Waldo Pepper, Waldo explains to a family about his encounter with the Red Baron during The Great War. The scene starts out in a fairly wide shot of Redford along with the members of a family in whose living room he is sitting. Then, as he tells the story of the twists and turns and machine gun fire, the camera slowly zooms in on Redford's face. The film could have called for a grand scene depicting the aerial dogfight, but the story told and the way it is set up in the motion picture makes the audience member feel like he or she is in that room with that family listening to the story. This suggestion of action and entrancement in the storyteller instead of the cheat of the visual flashback creates a relationship between the audience member and Waldo Pepper. This is far better than any visual aid. The audience understands the story regardless of the images provided, but now they are connected emotionally to the storyteller. The Action of the present plight of N_2 keeps rising in the sample presentation.

At this point the story turns to an unlikely hero. If the nitrogen molecule that makes up so much of our atmosphere cannot be observed, who will provide this useful information for its detection in Earth-like planets observed elsewhere in the galaxy? Enter the $N_2 - H - N_2{}^+$ proton-bound complex. He can serve as the representative for N_2 and to stand in its place for detection. Figure 4.17 introduces the principle protagonist of the story. He is described as being much like N_2 but with a unique trait. The slide now has the proton holding the two nitrogen moieties together animated in order to highlight this most important of vibrational modes. This cannot be shown in Figure 4.17, but the animation actually has the proton bouncing back and forth quickly between the two N_2s about a dozen times in five seconds.

An image then appears in Figure 4.18 to highlight that if 2% of the total mass is moving and roughly 100% of the charge, then the induced dipole will be bright. The fireworks photo (taken by the author at the July 4th celebration on the National Mall in Washington, DC) insinuates the intensity of the peak that should be observed in any vibrational spectrum of this molecule for this proton-shuttling mode. As a side note, removing the background in this

Proton-Bound Complexes

Figure 4.17 Rising action slide 6 stage 1.

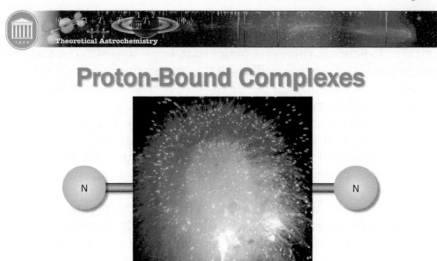

Figure 4.18   Rising action slide 6 stage 2.

photograph did not create a satisfying image. Hence, the colors are contrasted allowing for a darker (and, hence, forgettable but necessary) background. In any case, our hero has a unique character trait that will allow him to save the day. How he will use that trait is not entirely divulged yet.

Figure 4.19 contains a visual pun. Employing Beer's law ($A = \epsilon bc$) is necessary to detect this molecule. The symbols of the stein and gavel make the audience know that Beer's law is being employed without having to use text, and it makes for a nice chuckle. The fireworks represent ϵ. Figure 4.20 introduces c for concentration, again a visual pun that communicates the concept quickly and without text (also to a hopeful chuckle from the audience). However, Figure 4.21 shows the aftermath of an animation in which a clip art of Rodin's *The Thinker* is reduced in size. Hence, the reduction in concentration, c, combined with a large transition dipole moment, ϵ, implies that the absorption, A, can still be large enough to observe even if there exists very little of the $N_2 - H - N_2^+$ proton-bound complex in protoplanetary disks. Our hero has won his first battle and is charging to save our heroine of N_2.

Note that the narrative is not being told as a "knight-in-shining-armor" fairy tale. The data are being produced and the information given scientifically. However, the story is constructed in much the same way in order to paint a fireside story that has all the trappings of traditional oral presentation.

Figure 4.19   Rising action slide 6 stage 3.

Figure 4.20   Rising action slide 6 stage 4.

Proton-Bound Complexes

Figure 4.21   Rising action slide 6 stage 5.

Q. Yu, R. C. Fortenberry, and co., *J. Phys. Chem. A*, **119**, 11623, (2015).

Figure 4.22   Rising action slide 7 stage 1.

4.5.3   The False Climax and Continuation of the Rising Action

The narrative of our hero going into his first real battle is picked up here. The truth of the actual presentation is that the vibrational frequencies for this molecule are actually computed. This is initiated with a slide that first appears as Figure 4.22 and quickly transitions to Figure 4.23. The method described in Figure 4.10 does not work

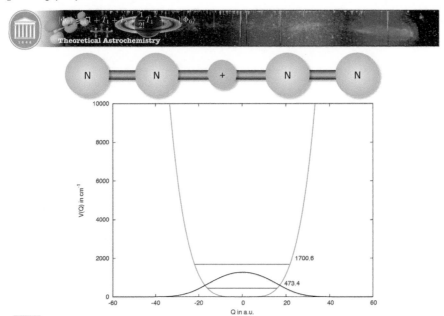

Q. Yu, R. C. Fortenberry, and co., *J. Phys. Chem. A*, **119**, 11623, (2015).

Figure 4.23   Rising action slide 7 stage 2.

due to a flat potential energy surface in the proton motion. The implication is that our hero has failed in our background narrative in his first major conflict. This introduces a struggle natural for him to overcome. However, through a research collaboration between this author and other quantum chemists, this vibrational frequency is computed (Figure 4.24), and the hero wins the day. This represents a climactic part of the narrative.

The story could end here. The data could be provided for ultimate use by infrared telescopes, and the Action could fall from the Climax into the Resolution. This would work very well for a 10–15 minute presentation.

Alternatively, this could be a false Climax, and the story could employ Rising Action, again. Such jumps up the Rising Action mountain often provide chances for the audience members to regroup and return to the story with renewed vigor. Most epics have such false Climaxes. *Beowulf* has three! Each progressive Climax, however, utilizes the previous one as a necessary stage for the story to continue.

In the present example, $N_2-H-N_2^+$ has a band of merry men to join in the search for interstellar N_2. The reality is that $OC-H-CO^+$,

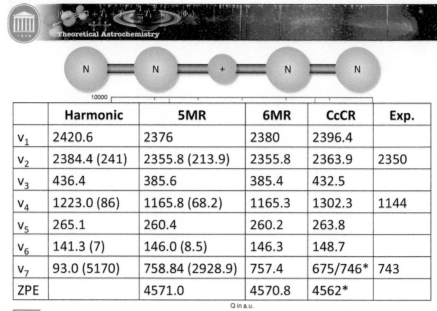

	Harmonic	5MR	6MR	CcCR	Exp.
v_1	2420.6	2376	2380	2396.4	
v_2	2384.4 (241)	2355.8 (213.9)	2355.8	2363.9	2350
v_3	436.4	385.6	385.4	432.5	
v_4	1223.0 (86)	1165.8 (68.2)	1165.3	1302.3	1144
v_5	265.1	260.4	260.2	263.8	
v_6	141.3 (7)	146.0 (8.5)	146.3	148.7	
v_7	93.0 (5170)	758.84 (2928.9)	757.4	675/746*	743
ZPE		4571.0	4570.8	4562*	

Q in a.u.

Q. Yu, R. C. Fortenberry, and co., *J. Phys. Chem. A*, **119**, 11623, (2015).

Figure 4.24   Rising action slide 7 stage 3.

$NN-HCO^+$, and $CO-HNN^+$ are introduced and their properties presented (not shown here).

In truth, even slides about the other three related complexes could even be left out as they are in this sample presentation. They are summarized in Figures 4.25–4.28 in a progression of stages but within a single slide. This cartoon shows where the bright infrared frequency will be for each molecule and how they relate to each other in terms of frequency/energy and intensity. These are introduced one at a time to give the audience a chance to distinguish the images and get a clear picture of how these different molecules relate to one another. In other words, the band of merry men is shown first by each individual's unique traits, and then together for comparison. Tolkien uses this technique in *The Lord of the Rings* by separating Frodo and Sam from Merry and Pippin. They each have their role to play in the story but cannot do so as a united group.

To continue our story, we give it a twist — that potentially this search for N_2 is not the most important thing. Rising action of a different vector is employed, but the two narratives will eventually meet, again, just like the two pairs of Hobbits. The confluence of multiple

Mass-57 Complex IR Spectra

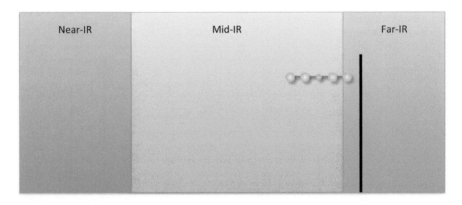

R. C. Fortenberry, J. S. Francisco, and T. J. Lee, *J. Phys. Chem. A*, **120**, 7745 (2016).

Figure 4.25 Rising action slide 8 stage 1.

Mass-57 Complex IR Spectra

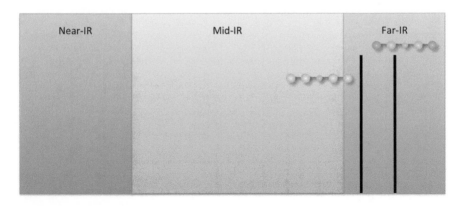

R. C. Fortenberry, J. S. Francisco, and T. J. Lee, *J. Phys. Chem. A*, **120**, 7745 (2016).

Figure 4.26 Rising action slide 8 stage 2.

Mass-57 Complex IR Spectra

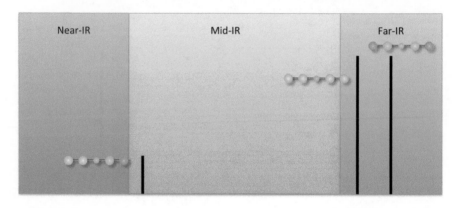

R. C. Fortenberry, J. S. Francisco, and T. J. Lee, *J. Phys. Chem. A*, **120**, 7745 (2016).

Figure 4.27   Rising action slide 8 stage 3.

Mass-57 Complex IR Spectra

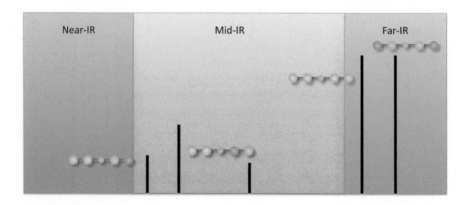

R. C. Fortenberry, J. S. Francisco, and T. J. Lee, *J. Phys. Chem. A*, **120**, 7745 (2016).

Figure 4.28   Rising action slide 8 stage 4.

narratives into a seamless tale is a time-honored technique and is not outside the realm of scientific presentation.

The narrative shifts here to describing the detection of naturally occurring noble gas molecules. Yes, noble gas atoms have chemistry in contradiction to their implied family name. Figure 4.29 returns to a common motif which is a good way to begin a second narrative without losing the audience. A new portion of the image motif is highlighted in Figure 4.30 to showcase the noble gas atoms involved. Then, Figure 4.31 has the stage of the slide where an image appears of the Crab nebula where, in 2013, argonium (ArH^+), shown in Figure 4.32, was discovered. These slides begin the path toward a new climax.

The credibility of the speaker, showing that he or she has worked in this field with much success, is established in Figure 4.33. The molecules are animated onto the slide individually, but that is not shown explicitly in this text. As well as confirming the speaker's credibility, this slide builds the Action by describing that there is more to noble gas chemistry than typically is described in the chemistry classroom. This is further enhanced by Figure 4.34 where even more molecules are shown to be plausible and their detectable vibrational frequencies are implied to have been provided in the literature. These

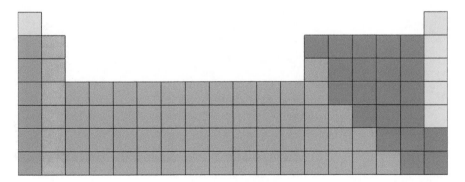

Figure 4.29   Rising action slide 9 stage 1.

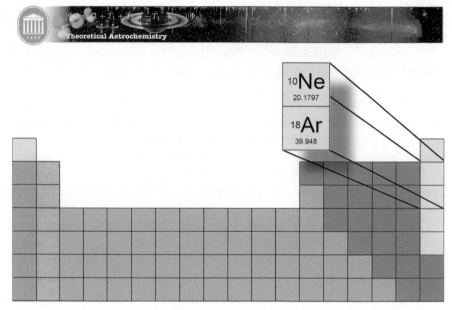

Figure 4.30   Rising action slide 9 stage 2.

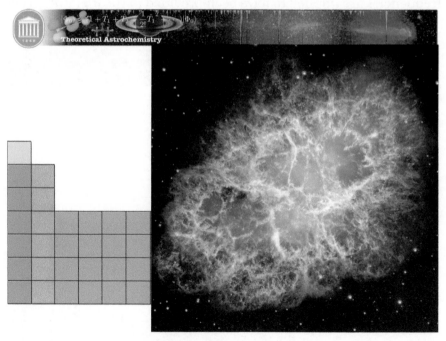

Figure 4.31   Rising action slide 9 stage 3.

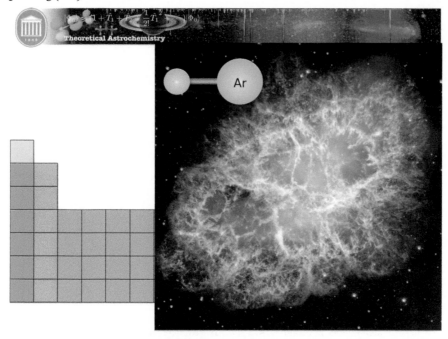

Figure 4.32 Rising action slide 9 stage 4.

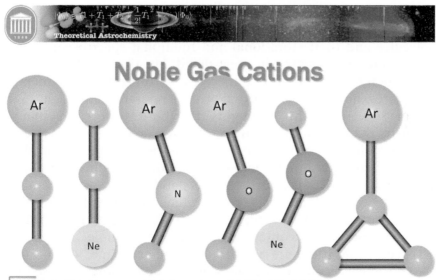

R. A. Theis, W. J. Morgan, and R. C. Fortenberry, *MNRAS*, **446**, 195 (2015).
R. A. Theis and R. C. Fortenberry, *JPCA*, **119**, 4915-4922 (2015).
R. A. Theis and R. C. Fortenberry, *Mol. Astrophys.*, **2**, 18-24 (2016).
C. M. Novak and R. C. Fortenberry, *J. Molec. Spectrosc.*, **322**, 29-32 (2016).

Figure 4.33 Rising action slide 10.

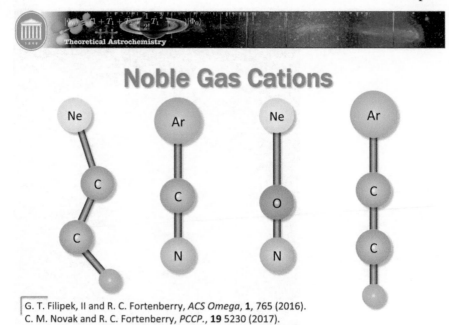

G. T. Filipek, II and R. C. Fortenberry, *ACS Omega*, **1**, 765 (2016).
C. M. Novak and R. C. Fortenberry, *PCCP.*, **19** 5230 (2017).
R. C. Fortenberry and S. R. Gwaltney, *ACS Earth Space Chem.* (2018), *submitted.*

Figure 4.34   Rising action slide 11.

molecules and their vibrational and rotational spectroscopic properties could be discussed in great detail, but that would only serve to lengthen the presentation unnecessarily. If the audience member really cares, he or she can write down the reference and look up the information. Again, this highlights that there is more to noble gas chemistry than is typically assumed. The Action continues to rise.

At this point, the two narratives reunite with one another in showing that there exist proton-bound cation complexes involving noble gas atoms. (Also, let the reader understand that in astrochemistry, elements heavier than iron are rare which is why only these three noble gas atoms are discussed; this is part of the live narrative.) Figure 4.35 showcases the six combinations of noble gas molecules possible in space. Again, these are animated in one at a time. Then, Figure 4.36 introduces numbers that are stated to be proportionate to bond strengths. Again, these numbers are faded in individually. Hence, noble gas molecules may yet be detectable in even larger numbers than previously thought. The Action is getting to a point of Climax.

The Action is just about to reach its ultimate peak. However, the vibrational frequencies for the proton-shuttle frequency are shown

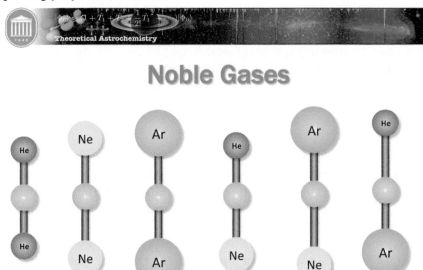

R. C. Fortenberry, *ACS Earth Space Chem.*, **1**, 60 (2016).

C. J. Stephen & R. C. Fortenberry, *MNRAS*, **469**, 339 (2017).

Figure 4.35 Rising action slide 12 stage 1.

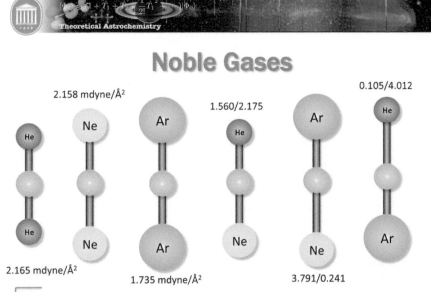

R. C. Fortenberry, *ACS Earth Space Chem.*, **1**, 60 (2016).

C. J. Stephen & R. C. Fortenberry, *MNRAS*, **469**, 339 (2017).

Figure 4.36 Rising action slide 12 stage 2.

in comparison to one another assuming the same concentrations in Figures 4.37–4.42. This is, again, the same motif used in Figures 4.25–4.28 making it a comfortable use of describing similar but unique information. This communicates that maybe our hero is not $N_2-H-N_2^+$ but could be one of these species. Which could it be? The stage for the climax is now set.

4.5.4 Climax and Falling Action

Just as a tangent line only intersects the function at exactly one point, the Climax is a fleeting moment of finality on which the entire story turns but turns on the proverbial, minuscule dime, often on the discussion of a single slide.

The Climax for this story is introduced in Figure 4.43 where, again, a previous image establishes a motif. Then, three of the molecules fall away in Figure 4.44 indicating that they will not be valid for astronomical searches. Figure 4.45 highlights that the most likely of these molecules to find is $ArHArH^+$. The bright light behind it emphasizes this point that will be described in the narrative. However, Figure 4.46

Noble Gases

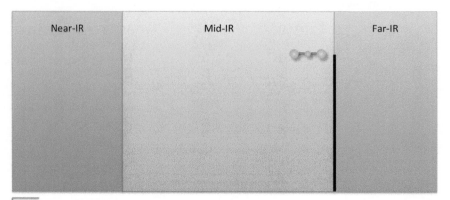

R. C. Fortenberry, *ACS Earth Space Chem.*, **1**, 60 (2016).

C. J. Stephen & R. C. Fortenberry, *MNRAS*, **469**, 339 (2017).

Figure 4.37   Rising action slide 13 stage 1.

Noble Gases

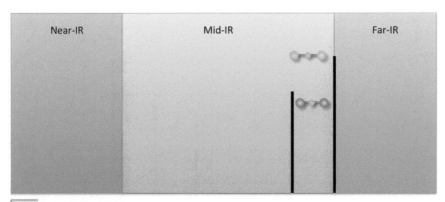

R. C. Fortenberry, *ACS Earth Space Chem.*, **1**, 60 (2016).
C. J. Stephen & R. C. Fortenberry, *MNRAS*, **469**, 339 (2017).

Figure 4.38 Rising action slide 13 stage 2.

Noble Gases

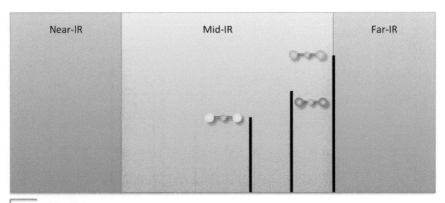

R. C. Fortenberry, *ACS Earth Space Chem.*, **1**, 60 (2016).
C. J. Stephen & R. C. Fortenberry, *MNRAS*, **469**, 339 (2017).

Figure 4.39 Rising action slide 13 stage 3.

Noble Gases

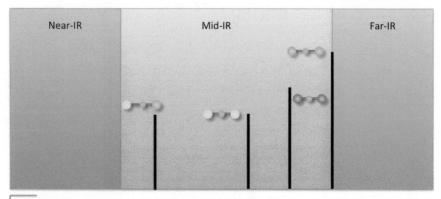

R. C. Fortenberry, *ACS Earth Space Chem.*, **1**, 60 (2016).

C. J. Stephen & R. C. Fortenberry, *MNRAS*, **469**, 339 (2017).

Figure 4.40   Rising action slide 13 stage 4.

Noble Gases

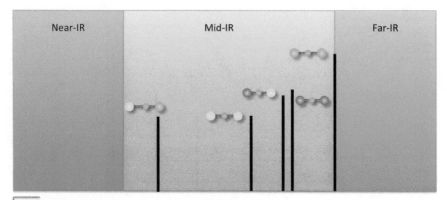

R. C. Fortenberry, *ACS Earth Space Chem.*, **1**, 60 (2016).

C. J. Stephen & R. C. Fortenberry, *MNRAS*, **469**, 339 (2017).

Figure 4.41   Rising action slide 13 stage 5.

Noble Gases

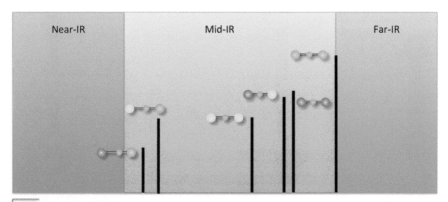

R. C. Fortenberry, *ACS Earth Space Chem.*, **1**, 60 (2016).

C. J. Stephen & R. C. Fortenberry, *MNRAS*, **469**, 339 (2017).

Figure 4.42   Rising action slide 12 stage 6.

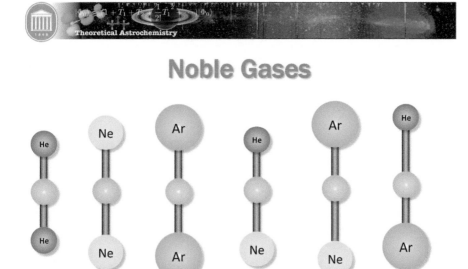

R. C. Fortenberry, *ACS Earth Space Chem.*, **1**, 60 (2016).

C. J. Stephen & R. C. Fortenberry, *MNRAS*, **469**, 339 (2017).

Figure 4.43   Climax slide stage 1.

Noble Gases

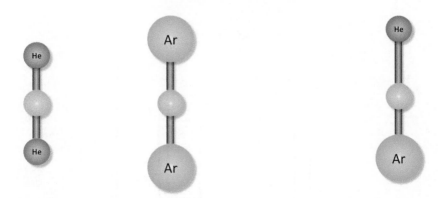

R. C. Fortenberry, *ACS Earth Space Chem.*, **1**, 60 (2016).
C. J. Stephen & R. C. Fortenberry, *MNRAS*, **469**, 339 (2017).

Figure 4.44 Climax slide stage 2.

Noble Gases

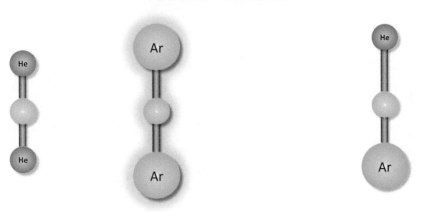

R. C. Fortenberry, *ACS Earth Space Chem.*, **1**, 60 (2016).
C. J. Stephen & R. C. Fortenberry, *MNRAS*, **469**, 339 (2017).

Figure 4.45 Climax slide stage 3.

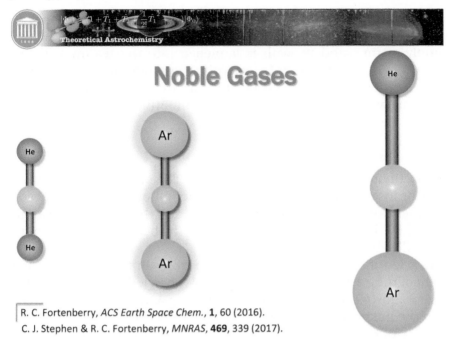

R. C. Fortenberry, *ACS Earth Space Chem.*, **1**, 60 (2016).

C. J. Stephen & R. C. Fortenberry, *MNRAS*, **469**, 339 (2017).

Figure 4.46   Climax slide stage 4.

shows that HeHAr$^+$ should not be forgotten for its rotational spectrum which is why it is emphasized as growing. Then, the HeHHe$^+$ is rotated around (not shown here) in order to indicate that it could still be detectable since there is so much helium present in space.

In truth, the earlier story of the nitrogen-containing proton-bound complexes was a bit of a red herring, but it established the discussion of these proton-bound complex molecules. That section had its own climax, but that was not the major and final takeaway of the narrative. The possible presence of ArHAr$^+$ is the most notable item from this story with minor contributors coming from HeHAr$^+$ and HeHHe$^+$.

This story combines findings from at least a dozen of the presenter's papers all woven together into a single narrative. This further emphasizes the discussion of the previous chapter in writing effective papers and having them published, and explains why the paper on HeHNg$^+$ molecules is used as the example. Regardless, the story in the presentation has now been told. The Falling Action is rather short and really begins once the Climax slide is transitioned from Figures 4.43 and 4.44. The Climax is infinitesimally brief and really is just the turn from the Rising Action into the Falling Action. The latter is simply tied up in Figures 4.45 and 4.46.

There is no Conclusions slide. There is nothing to state that the research was a success. It was neither successful or a failure. This is simply a story, and Beowulf is launched into the sea on a flaming funeral pyre. In the case of a scientific presentation, the story ends here in order for others to pick up the narrative and continue in much the same way that Frodo gets the ring from Bilbo.

4.5.5 Resolution and Ending

Often, the resolution is simply implied. The story is over. The hero wins or loses, but, in either case, satisfaction of a well-told narrative fills the audience. The campfire dwindles and the magical transportation of the listeners to a different land at the behest and words of the storyteller has drawn to an end.

In scientific presentations, the practical resolution is often a statement of future directions or simply the Acknowledgements slide. This author likes simply moving from the Falling Action statement directly into the Acknowledgements. Any future directions should be far too wonderful to be described since the reality of the future is hoped to be better than any speculations from the present. The students, mentors, and funding agencies are thanked. This is also a great place to put up pictures of the students or research group.

Then, the floor is opened for questions. Having a pertinent, funny, or beautiful image on the screen when the questions are solicited is a nice touch. Finally, the speaker is now charged with offering satisfying answers to questions raised by his story. These should be met with excitement as a means of continuing the story and not as dread for what he or she potentially did wrong in the minds of the audience members. Good questions beget a good relationship between the speaker and the audience. The answers should be concise and honest.

Ultimately, the narrative belongs to the speaker, and he or she is the medium of its conveyance.

4.6 Assignments

These are sample assignments for students to undertake. They build from watching others give presentations to having the students give such themselves. The students are expected to ask questions of their peers since communication should be a two-way conversation. While the presenter does most of the talking, the audience should feel encouraged, if not compelled, to ask questions of the presenter. One strong suggestion for the instructor is that he or she tally the number of verbal pauses ("um" and "uh") given by each student. This will make the student aware of such and work harder to minimize them.

4.6.1 TED Talks

The TED Conference was founded in 1984 and has since embraced online media as one of the most popular ways to disseminate the information gathered on technology, entertainment, and design. The leaders of this movement who promote "ideas worth spreading" have capitalized on the means and substance of effective yet informative storytelling. Some of their talks are the most viewed items on YouTube, for instance. Many of the speakers are coached before their presentations, but the ideas in the talks are all the presenters' own. Modeling their presentation style is not a bad way to proceed in giving oral presentations about science and chemistry, in particular.

Assignment for students: Watch two TED talks of your choosing on any subject. In a two-page paper, discuss the following:

- How long is the talk, and how does this influence your decision to watch the talk?
- With the time counter, determine how much time is devoted to showing the slides, and how much time is devoted to various angles and shots of the presenter? What does this say about the importance of the slides, and what does this say about the importance of the presenter.
- Comment on the placement of the speaker and how he or she utilizes his or her physical placement on the stage to build interest from the audience.
- Describe the stage setup and lighting. What moods does this promote? How does it make you feel about the speaker?
- Finally, what is the climactic point of the talk? What is the most important thing stated, and how is it developed?

4.6.2 First Presentation

Assignment for students: This is basically the oral counterpart to the first written assignment in Chapter 2. Go to an appropriate peer-reviewed, scholarly journal. Read and study the paper carefully. You will give a seven-minute presentation to the class allowing two minutes for questions. The instructor will give you 20 seconds leeway either way, but will reduce points beyond these limits. As a result, practice to yourself or a friend is valuable time spent. You must present the research in a clear way to the audience through scientific storytelling. Just like in writing, you must know your audience. You will *NOT* be utilizing a visual aid.

You will be graded based on the following criteria:

- 50% credit for simply performing the assignment within the allotted times.
- 20% credit for generating a talk that is relevant.
- 20% credit for the quality of the presentation both through the use of proper English and clear science journalistic communication.
- 10% credit for asking at least one question for two of your peers.

4.6.3   Second Presentation

Assignment for students: Building upon last week's assignment, you will give a similar seven-minute speech with two minutes for questions. You will be presenting a proper peer-reviewed, scholarly article from a journal, and it must be *different* from any of your previous submissions or presentations. This time, however, you will be constructing a professional visual aid including all of the necessary techniques discussed in class. You may use PowerPoint, Prezi, Keynote, or a similar professional style of modern visual aid. Glitter tri-fold posters are not allowed.

You will be graded based on the following criteria:

- Proper choice of topic (10%)
- Clear, concise, correct speaking (40%)
- Proper construction of data tables (10%)
- Proper use of figures (10%)
- Utilization of the visual aid beyond a means of displaying text (20%)
- Asking two questions of your peers (10%)

4.6.4   Final Presentation

Assignment (major) for students:

This major project will synthesize the previous two assignments but will be a presentation of your own work. Again, you will give a similar seven-minute speech with two minutes for questions. You will be presenting the research about which you wrote as your final assignment in the technical writing section. Again, you will be constructing a professional visual aid including all of the necessary techniques discussed in class. You may use PowerPoint, Prezi, Keynote, or a similar professional style of modern visual aid.

You will be graded based on the following criteria:

- Proper choice of topic (10%)
- Clear, concise, correct speaking (40%)
- Proper construction of data tables (10%)
- Proper use of figures (10%)
- Utilization of the visual aid beyond a means of displaying text (20%)
- Asking two questions of your peers (10%)

5 The More Common Presentation, the Poster

5.1 Introduction

A poster presentation is a chance to have a conversation (several actually) with interested passers-by. This presentation is for an audience of typically one or at most three at a time. There is dialogue, interruption, question, continuance, and, most of all, a chance to get to know one another well. Due to this dialogue and processing of information from another's perspective, poster sessions often leave the presenter with a deeper understanding of the material shown and research done. This is thanks to the insights gained from either having to explain the information in a novel way to a unique listener, or from novel interpretations from another individual with a different perspective. While some scoff at having to do posters, they are incredibly meaningful for those who do them. Granted, the number of people who see the presentation is almost always smaller than those who attend a talk or oral presentation, but those who do most often receive a deeper understanding of what is being described.

5.2 The Conversation

Just like with an oral presentation or talk, truly the most important thing in a poster presentation is still the speaker. The poster should be catchy and informative, but the printed material is still the visual aid. Just like the slides in a talk, the poster really should not be

Complete Science Communication: A Guide to Connecting with Scientists, Journalists and the Public
By Ryan C. Fortenberry
© Ryan C. Fortenberry 2019
Published by the Royal Society of Chemistry, www.rsc.org

able to stand on its own. The presenter has the job of telling a compelling story and utilizing the poster as a means of further expanding upon what his or her words cannot fully express. Hence, the poster presentation is still all about the person speaking.

When preparing for a poster, the presenter needs to create a spiel as an amalgamation of both storytelling and journalistic getting the most important information communicated first. There should be 30-second, 60-second, and the standard three-minute versions ready. Few things are more annoying to a scientist than having to listen to a researcher drone on about his or her poster, especially if that researcher is a judge for poster awards. Again, the old adage of "Stand up to be seen; speak up to be heard; and shut up to be appreciated," holds true. A ten-minute discourse shows enthusiasm and knowledge but often a lack of depth and insight. The presenter should never run the risk of boring the passer-by. This is a conversation, a dialogue, after all, and not a monologue. A complete but concise description is always best. If the curiosity of the passer-by is then piqued, he or she can ask questions at any point specifically to his or her interest. This opens up the conversation and tailors the discussion to the maximum amount of common ground between the poster presenter and the captive audience of one or two. Get to the point and be done with it.

The three-minute version of the story should be the most prepared. The plot diagram from the previous chapter still holds. This is a story that is being told, just in three minutes. Again, to use the example of getting lost on the way to a friend's house for a dinner party, the color of the street signs or the exact number of cars on the road is not important. The Setting of the story followed by highlights of the twists and turns in the Rising Action completed by the ultimate moment of Climax is what is most important. The presenter should never spend more than 30 seconds on Setting and should always give a clear marker as to when the Climax has happened. Non-verbal oral cues such as a slowing of the pace or a lowering of the voice are a great way to get this transition noticed. In all, the presenter should not get bogged down in the details. If the listener wants to know the details, he or she will ask. The presenter should give the high points in order to seem knowledgeable. This will then allow for a dialogue or simply for the passer-by to leave if he or she chooses.

The 60-second version of the presentation is really just an abstract. Often, the "short version" will be requested. Please, the presenter must provide this when asked to do so. A simple statement of motivation is provided, a brief précis of the work done is described, and the final result is given. The 30-second version is a shortening of this. The trick with these shorter spiels is to provide enough evidence that what is being said is credible and that there is a good more to be described if the passer-by only will ask. Truly, the most impressive thing to any listener of a poster presentation (or any scientific discussion for that matter) is if the story can be told completely and yet also succinctly. This demonstrates a rich depth of understanding and a command of the material. If it does not interest the passer-by then he or she can simply move on, and no one's time is wasted.

In any case, the poster is there for the presenter to utilize as a visual aid. The necessary items on the poster should be pointed out like they would be in a regular talk. The aid must still be utilized to inform about items when words would otherwise fall short. The presenter should also utilize non-verbal cues and body language to direct the passer-by. Eye contact is still vital as in any conversation, but the presenter should move his or her eye onto the poster when pointing things out in order for the passer-by to know to do the same. He or she will take cues from the presenter; the presenter must give them to him or her. The progressions should be made to be clear and meaningful, and the presenter should always be open to being interrupted. The best analogy is that the presenter is telling the passer-by about a hunt, to return to the primitive man example, with the poster serving as a map. Utilize the map and guide the singular audience member through the process as clearly and as invitingly as possible. Allow the listener to ask about certain features and items, and always remember that this presentation is about a story being told.

There are some individuals who do not wish to speak when they come up to a poster. They simply want to read the poster. Really, the presenter's job is to attempt to engage them anyway. The presenter should make eye contact with folks walking by, smile, and give off welcoming body language. Arms should not be crossed, and a pleasant demeanor (*i.e.* like you really, actually want to be there) should be communicated non-verbally. If someone stops at the poster, the presenter should ask a simple question such as, "What do you find interesting?" or, simply, "Can I tell you about my project?" Such questions initiated by the presenter can be welcoming to a stranger who may find the work interesting. However, some will simply say, "No." At this point, the presenter should simply smile, remain pleasant, and either say something like, "Well, just let me know if I can," or simply step back. The presenter should engage with the passer-by but should not make them feel cornered.

5.3 Cues from Art

In creating the actual poster, the presenter should keep in mind how human beings process visual information. The eye naturally flows in the English-speaking world from upper-left to lower-right. Part of this is because we read from left-to-right and top-to-bottom. This arbitrary choice has an affect on how information is processed.

Hence, the poster should have an opening for the eye to enter the scene in the upper-left-hand corner. The eye naturally wants to fall out through the lower-right. The poster should have visual items to stop this from happening. These cues have been developed by artists for centuries with the likes of da Vinci and Albert Bierstadt, the 19th-century German painter of the American West, being masters of manipulating the viewer into seeing only what the artists wants him or her to see.

A beautiful example of Bierstadt's work is part of the Smithsonian American Art Museum's collection and often hangs in the National Portrait Gallery in Washington, DC. "Among the Sierra Nevada, California" painted in 1868 is shown in Figure 5.1.

The brightly lit clouds ascending from the mountains invite the viewer into the picture of a truly breathtaking spectacle. The eye gets lost in the clouds and the mountains and either makes its way through the pass or along the mountain above the treetops to the

Figure 5.1 "Among the Sierra Nevada, California," Albert Bierstadt, 1868.

right. Either the rock patterns or the trees then bring the eye down to the lower-right where the shoreline directs the eye to a family of elk. The eye pauses here to take in this scene, and then, all of a sudden, the eye discovers a tiny waterfall hinted from the upturned heads of the elk. "Tiny" is of course a relative term since this painting is a whopping 1.8 m by 3.1 m. Now, the waterfall may have already been noticed, but now the eye is forced to contemplate this seemingly tiny yet beautiful item in this grandiose scene. The corners are all darkened, and the left-hand side is completely darkened. This keeps the eye returning to the bright sections of the painting right where the painter wants it.

Another example is "Landscape with the Fall of Icarus" by Pieter Bruegel the Elder in the Royal Museums of Fine Arts of Belgium. This painting is so moving that it inspired W. H. Auden's famous poem "Musée des Beaux-Arts" in 1938. In Figure 5.2, the eye enters the scene slowly and somewhat lost. The title says something about Icarus, the boy from Greek mythology whose father, Daedalus, fashioned him wings made with feathers held in place with wax. When Icarus flew too close to the sun against his father's warnings the wax melted, and the boy tumbled to the Earth to his death. This Flemish visual rendition seems to have nothing to do with Icarus. Again, the eye is lost. There is

Figure 5.2   "Landscape with the Fall of Icarus," Pieter Bruegel the Elder, undated.

a busy seaside town, a farmer is plowing his fields, a shepherd is tending his flock, and ships are sailing into port. Frantically, the viewer is forced to ask the question, "Where is Icarus?!" in a manner much like Daedalus would. Only when the eye follows the natural progression to the lower-right and the lines of the shore to an angler and/or the mast of the ship do two little legs appear in the water divulging the fate of Icarus.

Sure, not every viewer will find the same things interesting, but the cues provided in either scene have a general feel. Additionally, this makes for an event. While a photograph may be a memory, a painting is an experience that takes time to process. The brain has to engage and digest the material. Scientific posters must employ similar cues in order to engage the innate humanity for better communication of research findings.

5.4   Layout and Coloring

The most important portion of the poster is the middle. The upper-left invites the eye into the poster. The lower-right should stop it from going off the page. The lower-left has the least important information. The upper-right has the next-least important. Again, this is the way the eye naturally progresses. This was the staple layout in printed newspapers and magazines, but even many web pages have similar setups. While this overall trend of eye progression is true, the cues (pointing, looking at certain portions of the poster, *etc.*) from the presenter can change where the passer-by's focus lands, but these innate non-verbal processing cues are difficult to overcome. They must be utilized and not fought against.

The eye will follow any white space as a natural line of progression through the document. Having these well-placed and ending up at vital points or figures is a natural way to get the eye to fall on important information. Pictures, vertical/horizontal lines, or boxes are excellent at stopping the eye from moving too quickly off the page. Section headers can be used to stop the eye, especially when the information will largely be ignored by the viewer. Examples include "References" or "Acknowledgements." Such drive the eye back onto the poster and will not allow it to fall off.

The easiest thing to serve as a boundary is a darker area just like in Bierstadt's painting, Figure 5.1. Something else that makes the poster look darker is, in fact, text. Taken from a distance, black words on a white page (or any color page) serve only to darken the background. This wall of text will naturally be ignored by the viewer and poster presentation attendee. It will likely not be read except by only the most dedicated, interested, or introverted of researchers. Hence, text should be incorporated for only two reasons: (1) as filler, or (2) as a visual cue. Do not expect it to be read.

The use of color can also direct the eye in novel places. Furthermore, the choice of colors is imperative in a poster presentation. This affects layout and how attractive the visual aid will be to passers-by who could become the audience for a poster presentation. While such considerations are important for slides in an oral presentation, they are more important for a poster where the visual aid is the first thing those wandering past will see. The scheme of colors is employed in section-heading boxes, some text or text shadows, lines, or even the choice of colors for figures and plots.

The colors of text over background should always complement each other. Red should not fall on top of red or a shade of red such as pink or even purple. The colors will wash out and be hard to see. Having the proper contrast really makes the poster "pop," so to speak. This attractive feature will make posters more desirous for passers-by to stop and begin the conversation.

The color wheel is given in Figure 5.3. The three primary colors are red, yellow, and blue. Their wedges in the color wheel make a triangle. Posters utilizing these colors will have solid color contrast but,

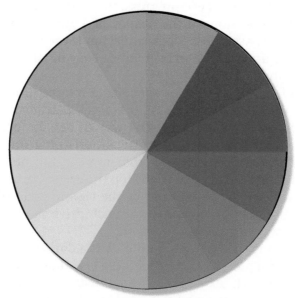

Figure 5.3   The color wheel.

frankly, will be boring. The secondary colors come from a 60° rotation of the color wedges to give orange, green, and purple. Posters utilizing these colors will stand out more and be less boring. However, overuse of such "unnatural" colors can become distracting quickly. The simple use of complimentary colors from diametrically opposed wedges in Figure 5.3 is a solid way to create pleasing color schema that do not distract. This is why the reds and greens of Christmas have worked so well for so long.

Another effective color scheme utilizes a trio of colors similar to the diametrically opposed version. Instead of selecting the single color immediately opposed the principle color, the two immediately beside the opposite color are chosen. For instance, blue-green with orange and red can make a useful association. There are other, more complicated color schema that can be employed, but keeping it basic while bright reduces the chance of distraction through color overload.

In truth, the choice of image, university colors, or necessary figure items will largely drive the choice of color beyond the point of the presenter's control. The key is to recognize what the major colors are in these required images. Then, utilizing a pair of colors that are either 180° or a trio of colors 120° from one another on the color wheel will make for a visually pleasing and stimulating showcase. For instance, if an image is taken of a forest at a lake's edge for inclusion in an environmental project, the natural blues and greens are complemented well by the rusty color of clay. Hence, if these colors are striking in the image, they can be utilized elsewhere in the poster to continue the motif. If they are already strong, then the complementary color(s) should be utilized, instead.

5.5   The Actual Poster

The size of the actual poster will vary by venue. The typical poster is roughly 1 m tall and 1.3 m wide (3 feet by 4 feet). However, some venues require portrait and not landscape (tall rather than wide) layouts for posters. In any case, the venue should supply the size of the poster area to be utilized before the poster session. The presenter should tailor his or her poster to the required size.

An early poster by this author is given in Figure 5.4. The sides of the poster are walled by text, and the eye is driven down into the poster by the sample molecules given at the top. The References are given in the lower-left since they will largely be ignored. Logos of the funding sources are given in the lower-right to stop the eye. The eye is driven into the middle of the poster where the results for this study are given, in this case electronic excitation energies for a family of molecules.

The largest issue with this poster is the amount of text. No one will read this, and any pertinent information from the text will be stated by the presenter. Hence, the text is useless. The data and images are useful and should have taken a larger role in the poster.

The best thing about this poster, however, are the colors. The spectrum of the diffuse interstellar bands, an unattributed interstellar absorption spectrum ubiquitously observed in most interstellar sightlines and known for over 100 years, is surreptitiously in the background. The section headings make use of this by brightening this background image as part of their need to stand out. The text of the section headings is actually the complementary color of the primary color in the heading box. Furthermore, the images of the molecular orbitals shown at the top of the poster match the color scheme of the underlying spectrum in the vertical columns beneath. These colors are attractive to passers-by and can get the conversation started. However, such excessive use of ink is expensive to print.

A later example of a poster by this author is in Figure 5.5. The protonated and hydrogenated forms of carbon dioxide are pertinent for studies of the atmospheres of both Mars and Earth. Hence, the theme and colors of the poster are rust-red and oceanic blue, which are on opposite sides of the color wheel in Figure 5.3. Additionally, the photos fading into the scene in the two upper corners makes for a nice transition without using too much ink. Even the layout of the title and the authors' names creates a "V" shape that direct the eye down

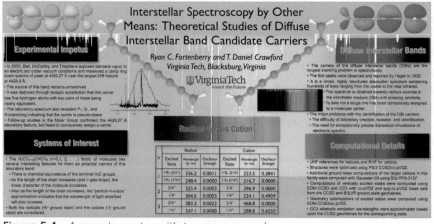

Figure 5.4   A sample poster with too many words.

and to the middle. There is significantly less text on this poster, and the counter symmetry of the complementary colors on both sides also makes for an attractive and striking poster in order to get the attention of passers-by. While there are walls of text on both sides, this is reduced in comparison to Figure 5.4 with more information in the form of data in the middle. The eye naturally falls to the *trans*-HOCO data in the left-center and then flows through the tables given. These data can then be described, pointed out, and developed by the presenter through conversations with the passer-by.

White space should be minimized on a poster, but, again, it can be utilized effectively to drive the eye to certain regions. Sometimes, this is why text is put on a poster; it simply takes up needed space. Truly, this author retains Abstracts on a poster placed in the upper-left hand corner and called "Introduction" most often. This author knows that such will not be read, but the gradation of color and the off-chance that someone may find this interesting allow for this text often to be retained. Still this is typically, again, just to take up space.

A student poster in Figure 5.6 has the sides walled in text, again, and has simple section headings. The Introduction/Abstract is in the

Figure 5.5 A sample poster with color and less text.

upper-left with Conclusions in the upper-right. The eyes flow naturally into the center to highlight the most important items of the spectroscopic data. There are a good number of images, even ones created by the student herself. This type of format is simple for students to emulate. This author encourages his students to begin with their posters like this. Then, they can try to get more creative. One student, at the end of her research career with this author, created the poster in Figure 5.7. This poster contains many images, data tables, and related figures. There are few words on the poster, lots of bright colors, and useful boxes around sets of information.

When creating a poster, the use of graphical editing software is best. Standard presentation software packages like Microsoft Power-Point and Apple Keynote really are the easiest. The slide size must be adjusted to fit the printer utilized, but these packages are relatively easy to use and quite powerful. Adobe InDesign can also be brought to bear but often requires a bit more expertise with the software. The use of mark-up software like LaTeX most often lends itself to over textualization of the poster and does not allow for proper construction of the poster layout as described above.

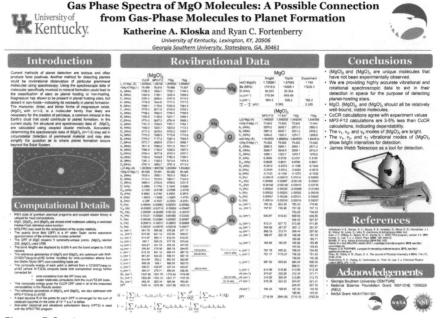

Figure 5.6 A more recent sample student poster.

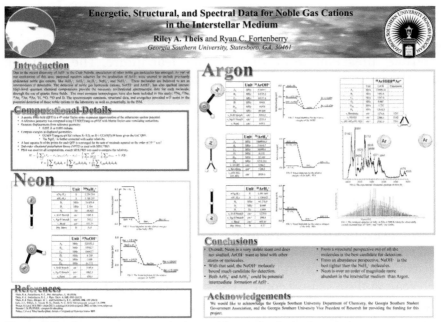

Figure 5.7 A more recent and advanced sample student poster.

As an anecdote, in this author's first year of graduate school, he made his first single-sheet poster. While in undergraduate research, the technology was changing, and he utilized a series of PowerPoint slides. However, the first conference at which he presented what is now the standard 1 m × 1.3 m poster came in May at the end of a year filled with growth. He finished his poster and showed it to his Ph.D. advisor. The advisor looked over the poster. He was pleased with many of the inclusions and discussions of both the poster and the way this author was going to present it. However, the advisor raised a question about layout. He said that he had never seen a poster arranged in such a way and did not feel this was proper. This author made an argument to keep it based on the tenets discussed above. At such a point the advisor, in his kind wisdom likely thinking this was either going to work or blow up in my face teaching me a lesson said, "Well, you have a cogent argument for making it this way, and it is your poster, not really mine. Keep in this format if you would like to." This author won an award for this poster at his first meeting which would turn out to be the first of many in graduate school. Hence, these concepts work, anecdotally, and, at the very least, will make the conversations at poster sessions more enjoyable for both the presenter and passer-by.

5.6   The Future of Poster Presentations

New technologies are emerging that will make the piece of paper in a "poster" presentation obsolete in the near future. The American Geophysical Union (AGU) has begun utilizing video screens instead of printed posters for a small subset of its poster sessions. As technology matures, this, or something like it, will likely become the future of *en masse* scientific presentations historically called "poster sessions," whether posters will be present or not.

The largest barrier at this time to electronic technologies implemented in poster sessions is cost. Providing thousands of monitors or even projectors for students and researchers to present their findings on a screen is greener in the sense that paper is not printed and then thrown away, but these screens have to be purchased, powered, and protected by security. This adds cost. Hence, the AGU is merely piloting such use. However, the price of flat-screen monitors is decreasing each year, and these may come to be cost effective at some point.

Even so, the most likely scenario is that researchers will travel with mini projectors that hook into their phones. These could be set up in a fashion similar to that done in traditional poster sessions, but the projection will take up the space on what would be the poster paper. Differently, flexible, organic-based electronic screens (potentially developed by one of you) may supersede all of these technologies where a mat is unfolded and hung up to be utilized for image presentation. Finally, immersive technologies may be better for communication where a virtual reality system is set up, and the passers-by and poster presenters interact virtually where the laws of physics are suspended within whatever virtual construct is best for communication. This can allow for novel interactions for the discussion of research findings in ways that cannot be imagined in the present age.

Regardless of the technology, the point is that in the relatively near future, printed, static posters will be a thing of the past just like transparencies, and slide projectors are now for oral presentations. This will allow for dynamic presentations at "poster" sessions where the tenets of the presentation described in the previous chapter will likely be more applicable than ever. The three-minute presentation will be more dynamic, but the "poster" presenter will have to go through this multiple times. Even so, having a series of slides is not likely to be the most effective means of communication since talking to one person via PowerPoint is often ungainly and awkward. However, dynamical visual aids will be employed where haptic manipulation of the visual aid will likely create new and creative means of communicating information. As such, new processes will be developed as this technology emerges.

While new technologies will develop, the most important thing to realize is that the presenter should always be the focus of any presentation, and the visual aid is simply an aid, not a replacement or a crutch. As scientists, we must embrace these technologies and utilize their nuances for communication. We cannot expect to simply transfer the old way of doing things into more modern media. For instance, having a static poster projected onto a screen would lose so much of what the medium can do for the sake of communication. The fullest extent of the technology must be brought to bear in order to have the most effective form of communication. Young people are almost always at the forefront of new technology, and they will lead the scientific community into avenues of communication. Adoption of new ideas while discarding those that do not work will ensure that scientists are always at the fore of information dissemination.

5.7 Final Considerations

These concepts for layout and color schema are not native to scientific presentation. They actually come from journalism and graphic design associated with newspaper printing, magazine layout, and advertising. Hence, these same tenets can be applied to the creation of notifications, newsletters, images, logos, websites, and even memes, especially for the promotion of anything including departmental functions, research findings, seminar announcements, and anything else involving marketing and public relations materials whose use is described in the next chapter. The trick with any graphic design layout is to be creative and try new things. Text is dull. A picture is truly worth a thousand words.

5.8 Assignments

5.8.1 Poster from the Literature

Assignment for students:

In a manner similar to assignments from the previous chapters, the student should take an article from the literature and, "simply," turn it into a poster. This could be the same article as that utilized to write a journalistic piece or the one on which the student gave a sample oral presentation. In any case, the figures and tables are already given. The student should take these items and transpose them onto the poster.

Data tables should be recreated in order to make them fit better on the poster where different text and size requirements will certainly be present.

The students will be graded upon:

- Layout (30%)
- Proper use of colors and color scheme (20%)
- Full use of the poster (20%)
- Incorporation of the data and figures (30%)

5.8.2   Poster Presentation

Assignment for students:

The student should take the previous assignment and actually give a poster presentation on the material. The instructor may choose to have this as a series of posters for judges or simply as a presentation to the class. In either case, the students should prepare a presentation of under three minutes to go along with the poster.

The students will be graded upon:

- The poster itself with the criteria from the previous assignment (50%)
- Staying within the time limit (20%)
- Effectively utilizing the visual aid to enhance the discussion for the presentation (20%)
- Exhibiting welcoming body language for the invitation to a discussion (10%)

5.8.3   Final Poster Presentation

Assignment (major) for students:

This poster presentation combines all aspects of the two previous assignments but asks the student to create a poster based on research that he or she has done, or in which he or she has participated. This should be done as "an actual poster session" where the students have their posters arranged for faculty, fellow students, and other interested parties to peruse. Other judges can be present if desired. The students should feel as though their work is being presented at a professional meeting and should take care to prepare in such a fashion.

The students will be graded upon:

- Layout (20%)
- Proper use of colors and color scheme (10%)
- Full use of the poster (10%)

- Incorporation of the data and figures (10%)
- Effective use of time to develop the narrative (20%)
- Effectively utilizing the visual aid to enhance the discussion for the presentation (20%)
- Exhibiting welcoming body language for the invitation to a discussion (10%)

6 Public Relations and Marketing, The Synthesis of Science Communication

6.1 Introduction: The Role of Public Relations for Science

Great discoveries require great press. Again, the Wright brothers almost did not get credit for the creation of the first heavier-than-air flying machine until the news of their discovery became fully public. The news cycle is built upon information. Television news anchors do not hear about things through telepathy; they have to be informed. In working for a company or organization where science research is being undertaken, technological development is ongoing, or applications of new discoveries are being harnessed to create new products, those whose role is to share such information have to be informed. While journalists can dig for information, most often, the organizations themselves are the first to alert the journalists as to what is happening. This bridge between the in-group who knows and the out-group who needs to know is the essence of public relations.

At this point in history, every private organization has a dedicated team of public relations (PR) and marketing professionals. Most often, the out-groups for such traditional means are buyers of the products that the company makes. However, universities and research organizations have been shifting their gears for their PR teams. While university athletics, enrollment, and student outcomes often are the main thrust of academic PR, research outcomes have really started

Complete Science Communication: A Guide to Connecting with Scientists, Journalists and the Public
By Ryan C. Fortenberry
© Ryan C. Fortenberry 2019
Published by the Royal Society of Chemistry, www.rsc.org

to become a mainstay of university public outreach. The reason is simple: good will. If a researcher at Northwest South-Central State University is part of a team that develops a treatment for Parkinson's disease, then this progress for the good of mankind will make NWSCSU a desirable place for students to enroll, alumni to donate funds, and talented researchers to want to work. This then feeds upon itself to create a seemingly sustaining system of financial security and growth for the university. As a result, science is becoming a major driver of academic politics. The modern scientist must be versed in this evolving aspect of their professional lives.

Differently, but no less importantly, the modern scientist is also becoming more and more responsible for developing an audience for his or her work. With so many talented individual researchers veritably flooding the world with new discoveries on a daily basis, many important things get lost. Sure, peer-reviewed publications are easy to follow with news alerts and searchable databases, but the number of journals is exploding, especially with the growth of open-access publications. As a result, many solid, important findings are getting overlooked and left behind. The scientist must make steps to be a PR practitioner for his or her own research. The dawn of social media has turned into mid-morning, and few scientists are effectively utilizing this relatively easy means of PR to promote their own work. While the university or organization employing the scientist can do some of this, the employed PR professionals are largely only concerned with major leaps in knowledge immediately applicable to the public. Hence, the regular, small steps of science must be celebrated by the individual scientist and shared in such a way that the information does not get buried in some esoteric journal.

In truth, scientists have known need to do this for centuries. This is why we have peer-reviewed journals and, most notably, conferences with oral and poster presentations. These are nineteenth-century means of PR. While there will never be a true replacement for speaking with another human being face-to-face, new technologies are allowing humanity to share ideas instantly and globally. There has never been such a time in human history. Even Ghenghis Khan's famously vast kingdom was dependent upon messages brought at the speed of a horse and rider. Now, when a bus fire in Durban, South Africa kills three and injures 15 others, the Inuit First Peoples of Canada can hear about it in a matter of seconds. The scientist must now utilize these same means of communication to promote his or her science and discoveries so that groups in disparate portions of the globe can collaborate more effectively and create new knowledge impossible to attain as separate groups.

While the second of these two motivating PR scenarios should never be the scientific researcher's principle occupation and focus, it cannot be neglected. Granted, this can be overdone. Not everyone on a certain Twitter feed needs to know how many mLs of solutions got poured into each trial flask, but they likely care when the final result of that experiment has been published indicating that a certain compound known to inhibit the growth of brain plaques in mice also does the same thing in pigs. Such feeds can also be picked up by other disseminators in order to broaden the audience further. The whole world may not need to the result of every study (and certainly no one needs to know the gritty details of every experiment), but scientific friends and collaborators may find this useful in some way. The onus now falls on the researcher to make them aware as quickly as possible. With modern technology, that awareness can be instant and should be as nearly immediate as possible.

PR has evolved with technology and always will. These practitioners are on the forefront for the uptake of new, promising communication technologies and methodologies. While scientists can rely upon the professionals for the big things, we must understand their jobs a bit better in order to help them do those jobs more effectively. Additionally, we must promote our own work when appropriate. Often, the best scientific results are not the ones initially and largely accepted. The results that receive the most attention are the ones that are remembered and utilized.

6.2   Publics and Audience

The first and most important item in any PR action or larger campaign is to establish the audience, the out-group that needs to know. The technical term for this is the "public" for whom the messages are tailored. The phrase "general public" is a bit of misnomer. Such a group is in no way uniform. In the context of scientific PR, "general public" often refers to those who are not scientists. However, even this is fraught. A biochemist and geochemist are definitely both chemists, but the discoveries by one in a certain field may be completely undecipherable to the other. Does this make one a "general public" for the other? Certainly not in comparison to a group that may also include trial lawyers, mortgage bankers, and professional football coaches. Hence, the "general public" is actually an association of smaller groups with fairly fluid borders where such groups are called "publics." Defining the target public is essential in effective PR at any level.

In many ways, defining the public is similar to defining the readers for a journalistic piece or the audience for a presentation. The choice of language and message is dependent upon such individuals who would comprise the desired group. However, in public relations, the definition of the public defines the means of communicating with them, harking back to Marshall McLuhan. Young people communicate differently than those of middle age. One culture will interact within its group in a unique manner when compared to another culture. The knowledge base and common ground of one public will be different from another. Defining these groups and the subsequent communication patterns within them is the essential first step in any public relations campaign.

The target audience has to be defined, and it must be defined as uniquely and clearly as possible. For instance, a chemical company that has developed a new coating for decreasing friction in aircraft control surfaces will limit the PR campaign regarding this new molecular species to those who need to know. This narrowly defined audience will likely include aerospace engineers, aircraft servicing managers, and the buyers/administrators who work for/with both. While such sales of the chemical can still be lucrative, the number of individuals who actually need to hear these tailored messages is quite small and narrowly targeted.

Consequently, the choice of the public can be made quite easily, especially in science. Most PR campaigns must gather information about demographics, socio-economic statuses, lifestyles, consumption patterns, preferred media, and similar details. However, scientific PR may often have many of these items already established. The esoteric nature of most science and technology has clearly defined publics as practitioners in the field. Even university recruiting is clearly defined in whom should be the recipients of the messages. Additionally, the age of self-selection in social media, such as followers and self-declared "likes," further reduces the need to research this information. Even so, having such clearly defined publics always allows for even more narrow definitions of the public to get the message out.

In any case, the public should be defined as narrowly as possible. This reduces the chances that messages will fall on deaf ears. Such sending of messages is a waste of time for all parties involved and contributes to sound pollution, spam emails, and many ills of modern living. A key to narrowing the audience or public as much as possible is simply to choose whom the audience is most likely to be. Then, this group can be further subdivided into other segments. Some can be ignored while others are vitally important. This process can be repeated several times in order for the final group to be as specialized as possible without becoming overspecialized. The narrowness threshold will be established in the "Research" portion of the PR plan described in the next section.

Like with many communication techniques described in this text, scientists have been employing a form of this audience-narrowing behavior for years by choosing where and how they publish their research articles. Specialized and even sub-specialized journals have existed for decades and even more than a century in some cases. Biochemists are typically not going to read about geochemical results in *Geochimica et Cosmochimica Acta* because the target audience for the journal has been narrowly established. This increases the potential impact of most research within the well-defined field (save for that of ultimately high significance). However, scientific researchers need to branch out beyond these traditional means of promoting their scientific findings in order to generate more exposure of their work. Again, the first step is in defining whom the audience should be and how to reach them most effectively.

6.3   The ROPE Method

One of the simplest means of developing a PR plan or larger campaign is the R-O-P-E method for Research, Objectives, Programming, and Evaluation. This simple mnemonic promoted by Prof. Jerry A. Hendrix in his series of texts, *Public Relations Cases* published by Thomson, contains all of the basic ideas of solid PR for any level and any purpose. This process starts with Research in order to get to know the target public after it has been uniquely defined.

6.3.1   Research

The principle question in the Research stage is simply: "What does my public already know?" The follow-up question is similar, "What would I like for them to find out?" This Research stage should be clearly distinguished from the scientific research being promoted or described in the PR campaign. The ROPE version of Research is to understand the target audience in order to create the most effective campaign possible. PR Research is a totally different item than the science from the laboratory.

The first of the two questions is the principle avenue of Research. Telling the target audience things they already know is a giant waste of time. Telling the target audience things beyond which they have the ability to grasp is a giant waste of time. Hence, this middle ground must be established. Part of this comes in determining the public so there is a bit of a feedback loop between audience research and defining the desired public similar to the Shannon–Weaver model. In any case, there are lots of methods for finding the answer to "What does my public already know?"

The standard form of answering such a question from a professional perspective is through quantitative research. Typically, a questionnaire is distributed to individuals who are believed to fall within the target public. The questionnaire or survey is constructed with both binary questions (yes/no or true/false) and continuum questions (on a scale of one to five…). The survey should have some demographic information on it and contain questions that will take a reasonably short time to complete. Five minutes is about the maximum that most individuals are willing to give to such a survey. Hence, less than 20 questions beyond the demographic portion is best. Additionally, these surveys must be kept anonymous.

Sample Survey. Please note that this survey is illustrating content and not format.
Demographic Information (please circle or fill in where appropriate):

1. Age:

   - 0–18
   - 18–24
   - 25–32
   - 33–48
   - 49–65
   - 66 and older

2. Sex:

   - Male
   - Female
   - Other:

3. Education Level:

   - Doctoral
   - Master's
   - Bachelor's
   - High School
   - Other:

Survey:

1. I get most of my news from which source:

    - Newspapers
    - Cable Television
    - Radio
    - Online News Websites
    - Social Media
    - Other:

2. When a news story about science comes over the air/is posted, I listen to it/read it:

    - True
    - False

3. Science news stories get me excited about new discoveries:

    - Strongly Agree
    - Agree
    - Neither Agree nor Disagree
    - Disagree
    - Strongly Disagree

4. I read scientific journal articles:

    - Daily
    - Weekly
    - Rarely
    - Never

While this above example is quite rudimentary, it shows the different ways in which questions can be asked. The questions of the same type should all come in sets, and, if there are continuum questions like those from question 3 above, either the emotion or the order should be reversed in alternating questions. This should be done in order to make sure that the respondent is actually reading the questions and not simply filling in mindlessly. In other words, if a positive emotion is utilized in one question, a negative emotion should be included in the phrasing of the next question. Alternatively, if

"Strongly Agree" comes first in the preceding question, "Strongly Disagree" should come first in the present question. The former method is better for varying the outcome and should be employed over the latter which can sometimes be confusing if the respondent is not reading carefully. Finally, survey questions should be short and *VERY* specific. Multiple questions, too many words, and other confusing verbosity should be absolutely minimized.

Another form of quantitative research comes in content analysis. This is simply measuring the amount of material that is being generated as a part of the whole. Content analysis must also be done with clearly defined terms and parameters. For instance, a simple content analysis could be the number of times that a particular friend on FaceBook posts articles relating to science out of the total number of articles posted. Another could be the number of press releases generated by a university that contain scientific laboratory research updates out of the number that are produced in general. Such data help to inform the PR practitioner about the portions and amounts of messages that are being communicated, either by the audience or by the disseminating organization. These metrics can be quite complicated and informative. Alternatively, they can be quite simple (depending upon the scope and amount of time invested in the PR campaign), such as the number of times that the "University of Nottingham" is mentioned per year in the online version of *The Sunday Times*.

Regardless, advanced content analysis and questionnaire data should really be conducted by professional organizations. Statistical analyses are necessary to get proper information from the data, and internal review board standards must be maintained. However, understanding this process can assist scientific researchers in understanding how the PR professionals with whom they interact will conduct their quantitative Research. Additionally, a simple content analysis from social media posts or the scientific literature can help to establish who may be a good audience for promoting one's own scientific research findings.

However, qualitative analysis is probably a better fit for the self-promoting scientist, but it definitely has a role in professional PR work. The simplest form of qualitative analysis is simply from interacting with individuals from the target public. Political pollsters and commercial advertisers do this regularly with focus groups. In these groups, individuals who meet the target audience criteria are gathered and asked to discuss various topics and provide feedback.

The PR professional takes notes and records ideas as to how to proceed.

In truth, scientists have also been doing this for quite some time. This is the essence of the question-and-answer time at the end of an oral presentation. An idea or set of ideas is proposed to an audience, and the audience responds. If the feedback is positive, the scientist can continue his or her work. If the response is negative, the scientific researcher may need to re-evaluate some content before moving to peer-reviewed publication. This process differs little from how PR practitioners lead focus groups, but the focus is different. For traditional scientists, the purpose is to refine the science. However, PR gets a better idea of how to promote the Objectives of the PR campaign.

Another form of focus group is simply to interact with members of the target public in an interpersonal way. Basically, go have a conversation with someone. This connection allows for dialogue and directed questions. Such an approach is time-consuming, but it creates a deeper understanding of a few members of the desired audience in ways that surveys and focus groups cannot. Often, such interactions are targeted at specific individuals with lots of power. These include CEOs, corporate buyers, or media editors. Again, this is similar to the recommendation that scientific researchers converse with program officers at granting agencies. These directed interactions with gatekeepers will allow for a clearer direction of promotion.

PR Research seeks to establish how to reach an audience and what needs to be done in order to reach that audience. Scientists must learn to offer feedback to PR practitioners within their corporate, academic, or government organizations in order to help define target publics and audience Research. Additionally, scientists should take advantage of these tools in order to know how to promote their own work more effectively.

6.3.2 Objectives

Like the target public, the Objectives need to be narrow, specific, and clear, including numbers, means, and a timeframe. They are also of the utmost importance because they drive the entire PR campaign. Additionally, they need to be realistic, reachable, and verifiable. They should also be phrased in the infinitive action form like "to change," "to increase," or "to develop." The two main types of Objectives are "output" Objectives and "impact" Objectives. The former is within the control of the PR professional. The latter is largely not. Any Objectives simply answer the following question: "What is hoped to be accomplished?"

Examples of infinitive-form Objectives include:

1. To increase sales of etching acid by 5% over the course of the next 18 months.
2. To produce four peer-reviewed publications in journals with impact factors of 2.5 or higher before tenure.
3. To have 1500 citations of my papers by the time I reach 40.
4. To increase graduate student enrollment within our department by 25% within the next 3 years.
5. To post three Tweets (one each for internal, external academic, and external alumni relations publics) for every publication produced by our department for the next 12 months.

Each of the above examples are clearly defined with very specific targets and times. Such specificity means that these Objectives can be clearly evaluated (in the "E" of ROPE described later) and likely refined for subsequent reintroduction. Perhaps graduate student enrollment increases of this nature are too lofty. Resetting the goal for the next three year period to 20% is more attainable. Alternatively, perhaps 1500 citations turns out to be too few. Increasing this to 6000 before the age of 50 is a better goal. In any case, the specificity of these goals provides for clearly defined tasks that serve as guides for the organization or individual setting them.

Output Objectives are fairly straightforward, and they follow Programming described next. Simply put, Objectives 2 and 5 from the above list are output Objectives. While there is some impact related to the type of journal desired, the scientist is responsible for producing such work. Within PR, however, most output Objectives relate to the number of Tweets, press releases, television appearances, events, and promotional talks given in support of the PR campaign. These most often fall under the guise of controlled media which is produced by the organization (Tweets and press releases, for example) and uncontrolled media which is produced outside of the organization (television appearances and talks, for example from the previous list). The controlled media output Objectives should *ALWAYS* be met, and those from uncontrolled media should be diligently pursued.

The impact Objectives are often the hardest to meet and evaluate because they involve the response of the target public. The hierarchy of impact Objectives is informational, attitudinal, and, finally, behavioral Objectives. In other words, behavior cannot change without a change in attitude which, in turn, cannot change without new information given to the target audience. While informational Objectives are often the easiest of the impact variety to meet, they are the foundation for the ultimately desired behavioral impacts.

Informational Objectives simply are goals to provide information to the desired audience. These can range from simple exposure to new ideas to full sets of instructions. They often overlap with output Objectives but are typically more specific about the content of the output. Growth of awareness for some issue or concept is a standard informational Objective.

Attitudinal Objectives are often the most difficult of the three to evaluate, but they are necessary for subsequent behavior change. Many times, attitudinal Objectives are simply to change a certain percentage of the target public's perspective on an issue or product. Most easily, a common attitudinal Objective within higher education would be: "To reinforce the favorable opinion of graduate teaching assistants by the dean's budgeting office during the spring semester." This example goal is simply to keep the status quo with respect to funding, but there is some underlying behavior hoped for that will accompany this attitudinal Objective.

The behavioral Objectives are easy to evaluate since the Objective within the timeframe either takes place or it does not. However, these are the most difficult to achieve since another individual or group must personally perform some action. The PR professional does not force or coerce the members of the target audience to do anything under modern ethical practices. They simply provide information and exposure to ideas so that the public's members can decide for themselves whether they want to engage in some new and different activity or not.

In order to bridge these three classes, consider the following three Objectives related to chemical pesticides, which have made modern agriculture possible but whose overuse can have negative impacts on the livelihood of the very farmers who use them.

1. To create 1000 FaceBook posts containing messages about the killing of honey bees, targeted at Yorkshire farmers' wives during the summer months of this coming year.
2. To decrease the favorable view of broad-spectrum pesticide use by 25% among Yorkshire farmers' wives during the upcoming planting and growing seasons.

3. To reduce the use of broad-spectrum pesticides by 10% in Yorkshire in next year's growing cycle.

Each of these Objectives follow the other from information to attitude and, ultimately, to behavior change. They are distinctly targeted
at achieving different goals, but each is dependent upon the previous. Hence, impact Objectives most often come in sets. There also
need not be a one-to-one correspondence between Objectives of each
type. For instance, there could a dozen informational Objectives and a
trio of attitudinal Objectives all for the sake of achieving one behavior
change. Typically in fact, such a pyramid is quite common.

6.3.3 Programming

Programming is the item within the ROPE PR model that requires
most of the time and effort. While Objectives are arguably the most
important item, Programming is the most necessary. Without doing
the actionable items determined in the previous step, any such plan
is worthless. The best laid plans of mice and men may go awry, but
the best plans left unexecuted are a total waste of time and effort. Programming basically asks: "What is going to be done?" This includes
any usage of media formats, special events, public speaking, and
interpersonal interaction that will be utilized to achieve the Objectives
provided.

The main purpose of the Programming is to achieve the informational, attitudinal, and behavioral Objectives laid forth in the previous
step. Just like with writing, either technical or journalistic, a theme
is a standard way to connect the various program actions together.
Often, this theme is simply the item being promoted itself. Political
campaigns for candidates in elected office are a classic example. However, most political campaigns have a slogan, color scheme, and logo
to distinguish themselves from the others. Like with anything in science communication, the theme of a science PR campaign should
be kept simple. The science is complicated enough. A theme simply could be the application of the scientific research whether for
"directly useful" purposes or more of "that's cool" impact, again,
depending upon the Objectives. PR professionals can elaborate these
ideas, and larger PR campaigns should have consultation from such
professionals. However, knowing that there needs to be a common
thread through all of the promotional Programming will allow the
scientist to interact with the PR professional in an effective feedback
manner in order to establish the best path forward.

The actions are the items to be done. Many of these are laid out in the informational or output Objectives. If a certain number of social media posts, advertisements, press releases, promotional videos, or similar media are to be output, the Programming simply follows suit. However, other events or actionable items may also need to be undertaken in order to achieve the Objectives. Before the advent of social media, many PR academics claimed that most of the world of PR was largely planning and executing events. Now, people from all over the world can gather for an event and not have to be even in the same time zone much less the same room. Regardless, there is still something to be said for genuine human interaction. Hence, planning some type of event, whether in-person or even virtual, is a major part of the PR campaign. Again, scientists have been hosting conferences and symposia for centuries, and these are quintessential PR events. Granted, such events are often split up into various different smaller pieces with different targets for specific research areas (*i.e.* publics), but science can be promoted through other means, especially in the industrial sector or even for student-recruiting/alumni-giving within a university.

Most of us in the modern world have attended gala events, benefit shows, fancy dinners, local festivals, and trade show expositions. These are typical PR events. The scope and scale of the event, again, largely depends upon the Objectives. For instance, celebration of the 100th peer-reviewed paper by a faculty member at a research university is definitely PR for the scientist's work. A dinner or reception is typical for such a celebration, and it makes others aware of his or her work. Related, but different, would be the groundbreaking ceremony for a new research building. This is much larger in scope since more publics would be involved besides those interested in the scientist or his or her research. Such an event is an excellent way to build community investment in addition to respect/appreciation for the science. Again, these depend upon the Objectives. Consequently, special events are a great way to build interest in the scientific research work done within any organization. Often, the event itself can have its own PR campaign. While promotion of any event is necessary, the event is not the ultimate goal of the campaign. These two things can be easily confused, but they must be distinguished.

In any case, the planning of an event or any further Objectives have to be achieved through effective communication. The target audience must be known in order to be impacted. Advertisements, flyers, social media posts, and the like are excellent and established means of information dissemination. Again, any printed or single-image media can follow the same tenets of construction as scientific posters discussed in the previous chapter.

Besides planning and executing events, the next most common task for PR professionals is the writing of press releases. These are simply articles written in typical news, journalistic style that are sent to media outlets. In preceding decades, they could be included in newspapers or magazines as copy that the printer did not have to pay a writer or journalist to create. They are written just like news articles as described in Chapter 2 of this text. The one caveat is that they often do not give credit to a particular author but are simply viewed as coming from a PR team. Sample press releases are given below in Section 6.5 of this chapter.

A press release is most often not the final product printed, posted online, or otherwise broadcast. Such a piece is typically utilized as an initial set of information for the journalist. Typically, this document lets the news outlet know about the item being publicized by the PR professionals and whether or not such an item is newsworthy. If the press release is deemed such, then the editor can assign a field journalist to explore the story more deeply, armed with the information from the press release. When the press release describes an upcoming event, for instance, the journalist can know what the event is about and who are the most important players at the event for subsequent interviews or photographs. Press releases can show up verbatim as copy, but they usually are the jumping-off point for the media outlet. Even though print copy is declining, press releases are still highly effective means of inviting traditional media sources like television news or even online news web pages to take part in a PR campaign.

Often, the most effective means of any attitudinal or behavior change comes from human interactions with others. These are becoming more mediated in the modern world, but valuing what someone else has to say is the most straightforward way of achieving any psychological change. Influential persons are the opinion leaders, and they will influence the larger group to which any individual attaches himself or herself. Scientists do this as sycophantically as any other human being. A Nobel laureate or the "father" of a certain field is often heralded as a final say in any scientific matter, wielding their immense credibility. Having the blessing of such a highly credible opinion leader is effective in achieving higher-order Objectives. This is why celebrity spokespeople are still utilized today. Such an opinion leader will influence a group which will influence the individual. This is why we all get excited if a renowned scientist in the given field cites our personal work.

The other side of the coin is simply to generate so many messages that exposure and presumed credibility is guaranteed. Such constant exposure will be remembered. Hence, it comes down to a choice between what communication practitioners call broadcast messaging and targeted messaging. Both can be effective; rarely can both be done. The choice between them, just like the choice of which medium to use or what the scale of any event should be, comes down to cost and budgeting.

The budget is often passed down from higher administrators, especially for departmental or even personal PR, but full-fledged PR departments will have their own budgets. Budgets are, of course, flexible, but this is often the most crucial part of any proposed PR campaign. Accurate estimates for any Programming is a must as this will highlight whether certain items are worth the cost for the possible impact. For instance, social media ads are relatively inexpensive and targeted whereas radio spots can be a bit more costly. However, the audience and target public will really dictate which gets selected. These estimates require a little bit of Research and can be incorporated as part of the initial phase of the ROPE method. Additionally, knowing the fine details of the event to be planned are also a must. Seemingly inane questions such as, "Do the napkins need to be printed with the corporate logo, or can they be plain?" are important and are answered simply by knowing the difference in cost. Events, in particular, need to be fully fleshed out and prices quoted before these are sent to superiors for approval.

In all the goal of Programming is to achieve the attitudinal and behavioral goals set forth in the Objectives, and this is most easily described as simply "getting the word out." Events are great, but they can be too costly for their potential impact. However, they can also generate lots of publicity. This is where effective Research is essential. Some of the best Research, for better or worse, is simply experience. Some things will be effective and some will not. Learning through Evaluation is essential.

As a final note of personal research PR promotion by the individual scientist, *DO IT!* This can be rather easily done, is always a good idea, and does not have to be difficult, expensive, or time-consuming. Putting a link to a newly published paper on a personal or even group/departmental research page is a good step in the right direction. However, this is still a fairly broad-range "shot-in-the-dark." There are other means that practitioners in a chosen field get their detailed information. Electronic newsletters, list-servs, discipline-specific websites/social media pages, and the like will be viewed by those in the field who care. Within astrochemistry, the AstroChemical Newsletter (http://acn.obs.u-bordeaux1.fr/) is a monthly emailed notification with abstracts submitted by researchers for their peers to read. The audience is self-selecting and is narrowly focused. Hence, papers, jobs openings, and conferences can be promoted directly to the field of astrochemists. Another example from computational chemistry is the "Computational Chemistry List, Ltd," (http://ccl.net/). While this is a website, updates about all things occurring in computational chemistry are updated regularly on its online message board making it a natural place to check for updates within the field. Most fields have such newsletters and websites.

A slightly different example is the FaceBook personality, Hashem Al-Ghaili. He regularly posts updates about many things in science that are fun or fascinating. Hence, the barrier to posting on his page is higher, but anyone can comment on his postings. Another example is the StarTalk blog. As a result, if a scientist has done research in a particular field related to a post on either of these examples or similar outlets, a comment with a link to a paper or, better yet, an explanation of the research can generate traffic to one's own personal scientific findings. There are entire organizations dedicated to helping others harness social media for their personal use so this present discussion will not go deeply that direction, but such effort is useful. In any case, the point is that finding the specific audience for any level of PR, whether personal or at the organizational level, has never been easier and can be done without a tremendous amount of effort.

6.3.4   Evaluation

Evaluation is the means of knowing whether the time, effort, and money spent on the PR campaign were worth the cost. Part of this process is an assessment of the success of the particular campaign, and, again, part of it is in establishing what worked and what did not with regards to the clearly stated Objectives. The Evaluation is both qualitative and quantitative. In truth, most PR campaigns are on-going. Consequently, the Evaluation stage for one cycle flows seamlessly into the Research stage of the next. Hence, Evaluation and Research are very similar but with slightly different goals.

Informational Objectives can be evaluated based on exposure and retention of the communicated PR messages. Exposure is relatively easy to assess in this age of click counters and website access counts. Retention, however, is necessary to engage attitude change in the next level of objectifying outcomes. In either case, most often the same tools as Research are necessary for assessment. These include the standard survey and focus group engagements discussed above.

Small-scale, non-professional, personal PR, however, simply comes down to behavioral Objective Evaluation. "Were the desired papers cited?" "Did the amount of grant monies increase?" These are straightforward to evaluate for the scientist. Larger scale PR campaigns can be as well. "Was the fundraising target met?" "Did the number of volunteers grow to the desired level?" In other words, "Were the behavioral Objectives met?" These can often just be observed as a matter of standard operation.

Some items require a little digging for Evaluation. For instance, the Objectives listed above regarding pesticide use in Yorkshire would require evaluators to survey the farmers about their pesticide use in the next growing season. Alternatively or additionally, pesticide sales could be quantified as could the amount of pesticide run-off gathered from a local stream. In such an example, science and PR are intricately related in the Research/Evaluation steps.

Regardless, behaviors will likely only change once attitudes change, which will only change once the exposed information is retained. Hence, PR on the organizational level will require data about the attitudinal Objectives. Again, such analyses are best left to professionals, but understanding this process will make the professional scientist a valuable resource for the PR practitioners within the organization.

6.4   The Public Relations Plan

Any PR campaign requires a detailed plan. This document houses all of the necessary procedures for Research; clearly defined Objectives; all Programming including press releases, scripts for radio spots or online videos, copies of visual advertisements or flyers, samples of Tweets, a timeline of the campaign, *etc.* as well as the budget; and, finally, how the campaign will be Evaluated including how success will be gauged.

With regards to science and scientific research within a PR campaign, the PR plan really is the synthesis of this entire text. Typically, the practicing scientist will be involved in PR once he or she makes a novel, exciting, and otherwise noteworthy discovery or advance. The scientist's employing organization would likely be very keen to capitalize on the scientific researcher's findings. Hence, this discovery will be promoted by the organizational PR department, but the scientist should involve himself or herself in the process in order to provide an objective evaluation of how the promotion is being done.

Additionally, such promotion will likely involve the researcher for interviews, promotional videos, press releases, and the like.

Typically, the first stage in any science-related PR campaign is the acceptance of the seminal paper. The PR usually centers around the discovery and is timed to be at its peak on the date of the publication in the peer-reviewed journal. Hence, the technical writing always comes first. Then, the journalistic writing and information communication takes place. Again, these can be press releases written like journalistic news articles, Tweets, blogs, and even organizational website media. Any events will likely require the scientist to present and give a talk of some sort.

Consequently, the entirety of this text will be executed in a PR campaign. Such campaigns, whether on the small- or large-scale, are required to advance scientific careers. Hence, the scientist most often must be his or her own PR firm, even if the target public is none other than his or her professional colleagues or the organizational PR department at his or her place of employment.

This author has implemented such a style in the promotion of his own science in order to advance his career. He writes as many scientific papers as possible, gives as many talks as possible, and has clearly defined objectives and goals leading toward an overall personal PR plan. While such an approach will vary from researcher to researcher, having a PR mindset for science will open many doors for advancement that would otherwise be closed.

6.5   Sample Press Releases

6.5.1   Nitrogen Grabbed

[*For immediate dissemination to media sources in the Free State of Bavaria and to science news sources worldwide.*]

   [*Title*]

Newly Designed Molecule Binds Nitrogen

Universität Würzburg Press Release, February 22, 2018

   [*Begin text*]

Whether wheat, millet or maize: They all need nitrogen to grow. Fertilisers therefore contain large amounts of nitrogenous compounds which are usually synthesised by converting nitrogen to ammonia in the industrial Haber–Bosch process, named after its inventors. This technology is credited with feeding up to half of the present world population.

Air consists of nearly 80 percent nitrogen (N_2) which is, however, extremely unreactive, because the bond between the two nitrogen atoms is very stable. The Haber–Bosch process breaks this bond, converting nitrogen to ammonia (NH_3) which can be taken up and used by plants. This step requires very high pressures and temperatures and is so energy intensive that it is estimated to consume 1% of the primary energy generated globally.

"So we were looking for a way to split nitrogen that is more energetically favourable," explains Professor Holger Braunschweig from the Institute of Inorganic Chemistry at Julius-Maximilians-Universität Würzburg (JMU) in Bavaria, Germany. Certain bacteria show that this actually works: They are capable of doing so at normal pressure and temperatures by using the nitrogenase enzyme which catalyses the reaction with the help of the transition metals iron and molybdenum.

"We have been unsuccessful in reproducing a kind of nitrogenase so far," Braunschweig says. "So we started to look for an alternative: a molecule that is capable of catalysing the reaction and is not based on transition metals."

His team has been studying specific boron-containing compounds, the so-called borylenes, for years. They are considered potential candidates for such a catalyst. But how exactly would the corresponding borylene molecule have to be structured for this purpose?

The iron and molybdenum in the nitrogenase are known to give away electrons to the nitrogen molecule, a process called reduction. This causes the bond between the two N atoms to break. However, this only works because the transition metals are a good match for the nitrogen molecule: Their orbitals, the space where the electrons passed during reduction can be found, overlap considerably with those of the nitrogen due to their spatial layout.

Based on quantum mechanical predictions, Dr. Marc-André Légaré from the Institute of Inorganic Chemistry designed a borylene with a similar orbital arrangement. The results of his investigations were then synthetically tested at the JMU institute.

And successfully so, as the borylene produced in this manner was capable of fixing nitrogen – and that at room temperature and normal air pressure. "For the first time, we were able to demonstrate that non-metallic compounds are also capable of accomplishing this step," Légaré emphasises.

However, this does not mean that the Haber–Bosch process is about to be abolished. For one thing, it is not certain that the reduced nitrogen can be detached from the borylene without destroying it. However, this step is necessary to recycle the catalyst so that it is available to bond to the next nitrogen molecule subsequently.

"Whether this will ultimately yield a method that is more favourable energetically is still an open question," says Professor Braunschweig. "It is only the very first step, albeit a major one, on the way to reaching the ultimate goal."

The results of the study, which was carried out in collaboration with the research group of Professor Bernd Engels of the JMU Institute for Physical and Theoretical Chemistry, will be published in the renowned *Science* magazine.

[*end*]

6.5.2 Anions in Space

Note: This piece was written by this author and submitted to the NASA Ames Research Center Office of Public Relations. That office chose not to send out the story, but this serves as a model for professional press release style scientific writing in non-academic PR settings. Note the similarity to a typical journalistic news piece.

[***For immediate dissemination to media sources in the San Francisco Bay Area and to science news sources worldwide.***]

[*Title*]

Getting a Charge Out of Space

[*Begin text*]

Mountain View, Calif. – Theoretical chemists at the NASA Ames Research Center and SETI Institute recently matched spectral lines observed in the Horsehead Nebula to the C_3H^- anion. The astronomical region where the radio telescopes originally observed these lines is known to be filled with UV light, the same type of radiation that leads to sun-burned human lobsters after a day at the beach here on Earth. This so-called photodissociation region (referred to as a PDR by astronomers) was long thought to be too harsh for anions to exist. However, this new finding is challenging that long-held belief.

These rotational spectral lines were originally attributed to C_3H^+, a cation, but the data necessary to confirm this original attribution is lacking. In fact, this same group comprised of Drs. Xinchuan Huang, Ryan C. Fortenberry, and Timothy J. Lee were the first to question the association of these spectral lines to C_3H^+ based on their calculations. Huang says, "Our calculated rotational constants matched those observed in space very well, but we failed to match an important spectroscopic constant, which is worrying and makes the assignment questionable." In order for a molecule to be confirmed in space, all of the spectral features must agree.

Dr. Michael C. McCarthy of the Harvard-Smithsonian Center for Astrophysics in Cambridge, Mass. was the first to suggest C_3H^- as the molecule responsible for these lines since, as stated by Lee, "The spectra dictates that the molecule must have all of its electrons in pairs and be structurally similar to C_3H^+ without actually being the cation." Only C_3H^- matches these criteria, even though contemporary wisdom says the UV light is too harsh for it to exist in a PDR.

However, Dr. Martin Cordiner of the NASA Goddard Space Flight Center in Greenbelt, Md. says, "Studies of anions are still in their infancy... Currently, our understanding of anion chemistry is impeded by a lack of knowledge of anion formation and reaction mechanisms in the very low density environments found in space." Hence, the possible presence of C_3H^- in the Horsehead Nebula PDR forces astronomers "to go back to the drawing board to come up with a new way of understanding how anions form and behave in PDRs."

According to their paper set to come out in the *Astrophysical Journal* on July 10 (2013), the Ames group is not saying that C_3H^- exists for long in the PDR. However, enough C_3H^- exists at any given time for it to be detected. UV light in a PDR rips electrons off various atoms and molecules, but these free electrons can attach quickly to the neutral C_3H radical, already known to exist in this region, in what is called a dipole-bound excited state. The anion relaxes to its ground state, via radiative relaxation, where its rotational signal can be picked up by astronomers. Shortly thereafter, C_3H^- is destroyed through collisions with other molecules or more likely by the extra electron being ripped off by UV light. Dipole-bound excited states of anions have not been included in models that describe the chemistry of space, and this alternative attachment pathway of an electron being drawn in by the dipole moment of the neutral C_3H radical represents a new direction in interstellar anion formation theory.

Theoretical chemistry has already shown that C_3H^- not only has a dipole-bound excited state, but also a rare valence excited state. These theoretical chemists are now suggesting that taking into account electronically excited states, especially dipole-bound states, may be necessary in the formation of anions in space. Additionally, this study represents the first time that theoretical chemists are reassigning astronomical lines with no laboratory data. According to Fortenberry, "The computational methods and techniques have reached a point where theory can stand on its own... in the right circumstances."

[*end*]

6.6   Assignments

6.6.1   Annotated Event Budget

Assignment for students:

Imagine that you are working for your university department where you are currently enrolled. The chair/head of the department wants to increase the quality of student applications for undergraduate admissions. You need to plan an event, but, more importantly, you need to promote an event that you feel will attract students to your department. This will not be a paper, but you will need to write this in such a fashion that it would be submitted to the chair for his or her consideration. You have a budget of $5000. You will need to include a detailed budget (what and how much), a promotion timeline (when), a list and description of activities leading up to and on the day of the event (what and whom), and an idea of how you would generate novel ideas for student interest (the R portion of ROPE).

The students will be graded upon:

- Choice of activities/programming (30%)
- Effective use of funds (20%)
- Effective explanation of funds (20%)
- Clear descriptions of all items involved (30%)

6.6.2   Public Relations Plan

Assignment (major) for students:

You are to produce a public relations campaign in its entirety. Choose a scientific research entity of your choosing. This could be a university, government lab, private corporation, or anything that sponsors fundamental research and development. You are promoting a major scientific breakthrough. This can be purely fictional, but it should be realistic. For instance, we do not need to be curing all of cancer or finding intelligent aliens with whom we can discuss philosophy.

Have a little fun with it, but you need to properly follow the ROPE method. Again, you will need a similar document as used in last week's assignment with a detailed budget (what and how much), a promotion timeline (when), a list and description of all activities promoting the discovery, and an idea of how you would generate novel means of promoting this. Additionally, you will include any and all press releases (at least one), Tweets, Facebook posts, descriptions of photographs to be made, figures for promotion, AND a genuine research article for submission to *Chem. Sci.* Yes, you need to create a real journal article and its promoting press release. This will be a fairly large document with all of these pieces. You will also be giving a 20-minute presentation of this PR plan utilizing a visual aid on the day of the final exam.

The students will be graded upon:

- Choice of topic (5%)
- Press release (10%)
- PR plan and other examples (25%)
- *Chem. Sci.* article (30%)
- Presentation (30%)

Note that each of the subsections will be graded using the criteria provided for each of those previous assignments

6.7   Sample Public Relations Plan

The following is a sample student public relations plan reproduced with permission from the author.

2018

Kicking MBD1 to the Curb

Public Relations Plan by:

Shelby Lee Scherer

5/2/2018

Table of Contents

Introduction

Mission Statement:

The focus of this campaign is to increase collaboration and awareness of MBD proteins and other transcription factors to determine ways of minimizing the negative effects.

Problem Statement:

Kicking MBD1 to the Curb is a fundraising and enlightening experience for scientists and students alike to expand their knowledge of transcription factors, specifically MBD1, enjoy a fundraising experience, and the joy of soccer game The broad group of individuals enticed by this event will hopefully lead to further collaboration and break-throughs with the MDB1 protein and other transcription factors that bind methylated DNA

Research
Situation Analysis

Kicking MBD1 to the Curb is a weekend long event put on by Shelby Scherer and her research group at Georgia Southern University. This research group is a group committing their time to exploring the MBD1 protein in a research lab at Georgia Southern University in Statesboro, Georgia. The research is centered around the synthesis of different combinations of the MBD1 protein binding sequence to determine a peptide that is more specific to the methylated DNA. This project aims to prevent the binding of MBD1 which will eliminate the proteins function of preventing the transcription of tumor suppressors.

Participants will begin the weekend with drinks and dinner where executive director, Shelby Scherer, will explain the purpose and importance of this research breakthrough. A few big named scientists such as Dr. Mark Searle from Nottingham, England will be speaking. This will be followed by dessert where they will have the opportunity to engage with others about their own research. However, this is just the beginning. Saturday and Sunday will follow with an exciting 6 vs 6 soccer tournament that all are welcome to join as a player or to watch. Championship team will receive $750 grant towards research and the runner up will receive a $250 grant towards research. All proceeds will go to further research.

Other symposiums and research conferences are available to scientists. The goal of this conference is to bring experts together to not only learn more about each other's research and collaborate but to have a little fun kicking a ball around. The event is aimed to be intuitive, inspirational, and enjoyable.

The event will help fund Shelby Scherer's research and bring people together. Additionally, the event will help the students of Georgia Southern University obtain important people skills and volunteer hours. The presence of Tormenta FC, Statesboro's own minor league soccer team will help bring in the community to get an idea of the big things happening at Georgia Southern. The larger presence of local groups and companies could lead to additional funding and sponsorships for research.

The primary target audience are professors and research groups with a biochemistry background, focusing on transcriptional regulators and their binding to methylated DNA. The budget is based on the payment to attend the seminars and to participate in the tournament.

The aim is to expand the general knowledge of the MBD1 protein and the great breakthroughs that have followed and raise money for future research.

Organizational Analysis

Internal
Strengths:

- International/Size: Biochemists are all over the World and this would provide a meeting point to converse about similar research
- Dynamic: This event encompasses science and activity to keep participants active and motivated
- Small time commitment: The event is only a weekend so many students and professors would not need to worry about missing class or work
- Unique: Many conferences are centered around only talks and presentations. Kicking MBD1 to the curb is encompassing presentations and talks as well as some friendly competition.

Weaknesses:

- Expensive: Multiple venues and catering for multiple days will be reserved
- Worldwide: difficult to research everyone around the world and for them to come to a weekend long event.

External
Opportunities:

- Growth: many of the goods are insourced by Georgia Southern increasing their revenue and volunteer hours. The community such as Tormenta and small businesses are asked to join and buy banners as sponsorship to engage the community
- Collaboration: providing a great atmosphere for scientists with similar concentrations to discuss

Threat:

- Crisis: *Kicking MBD1 to the Curb* do not have a plan ready if a crisis were to occur
- Unnerving individuals with lack of invitations: the main campaign uses the internet, and some may not like the informality of it

Target Audience

Primary:

The primary audience are the head researchers and graduate students in labs around the world. The primary audience is specifically researchers between 18-65 who specifically study transcription factors that bind methylated DNA. More specifically, scientists that have a focus in the MBD proteins. This would promote the highest common knowledge and interest.

Secondary:

The secondary audience will be students from all over the world looking at presenting their research and playing soccer. Other students will also be important to help run the event through volunteering.

Opinion Influencers:

Opinion leaders will also be important in order to reach all of the masses. This group includes professors, students, clubs, bloggers, journals and any other form of creating awareness of Kicking MBD1 to the Curb.

Goals and Objectives

Goal: The goal of this event is to expand the awareness of methylated CpG-binding domain proteins and like research from around the World.

Impact Objectives:

1. To have at least 250 people attend the 'Kicking MBD1 to the Curb'
2. To have at least 8 soccer teams of participants for 6 vs. 6

Output Objectives

1. To post 3 Facebook status' a week to promote Kicking MBD1 to the Curb
2. To have at least two press releases printed about the MBD1 publication or the event

Informational Objectives:

1. To publish 10 volunteer ads in Georgia Southern's George Anne to get at least 50 volunteers

Attitudinal Objectives:

1. To increase the number of individuals who choose to 'attend another conference' in the post survey 20% from those who said 'would not attend another conference' in the beginning survey.

Behavioral Objectives:

1. To have 75% of individuals respond to the post-survey that the conference 'exceeded expectations'

Programming

Key Messages

Theme: "Kicking MBD1 to the Curb"
Messages:

- Kicking MBD1 to the Curb is a weekend long event to share research perspectives and partake in a soccer tournament.
- Kicking MBD1 to the Curb will occur October 12th-14th.
- The invitation for the event will be targeted towards Professors and Graduate students working with transcription factors binding methylated DNA, more specifically MBD proteins.
- This weekend long event will include check-in, dinner, and entertainment Friday night. Saturday will include a light breakfast and morning seminar, day-time soccer games, dinner and night-time seminar. Sunday will include a light breakfast another seminar and day-time soccer games.

Channels:
Targeted Marketing, Networking with previous event holders, PSAs, News Media, Georgia Anne at Georgia Southern and sister collegiate papers, and Facebook

Strategies

- Send press releases to *Georgia Anne* at Georgia Southern and have them send the story to other school across the Country
- Promote "Kicking MBD1 to the Curb" with Facebook ads and posting on the ACS Website
- Set press releases to be Shareable on Facebook to broaden scope
- Attend Club Sports meetings at Georgia Southern's campus and post help wanted ads in the Georgia Anne for volunteers
- Send official invitations to Dr. Mark Searle and Dr. Peter Dervan and then have follow up phone calls and emails

Tactics

- Print Media: Create press releases for college and university new papers
- Logo: Use for t-shirts to sell for awareness and to raise funds, on advertisements, and registration broaden outreach
- Social Media: Create social media content leading up to the event. Facebook is the primary choice to reach the target audience. ACS journals and newsletters will also include advertisements.

Timeline

2018
May:

- Prepare logos
- Create social media and advertisement channels
- Send press release for article to Georgia Anne to be sent to other Universities for publication on May 25th. Also send to ACS for publication
- Network with ACS to find biochem 'big wigs' to present
- Discuss the event with Georgia Southern Campus, Recreation, and Intramurals (CRI)
- Reserve RAC fields on 25Live
- Reserve PAC for October 12th-14th
- Create volunteer set-up and begin recruitment for food set-up, refereeing and other services
- Go to Georgia Southern's final club council meeting to discuss volunteering with club sports
- May 26th, paper is released advertising begins on facebook
- Press release is released for paper

June:

- Continue networking with ACS
- Send out initial emails to promote interest and awareness
- Send official invitations to Dr. Dervan and Dr. Searle
- Begin posting advertisements on Facebook and other social media
- Confirm dates, times, and requirement needed with CRI
- Ask Tormenta FC and local companies about sponsorship or volunteering
- Begin Registration for posters and seminars
- Press release for event is released, send to Georgia Anne to be sent to other Universities.
- Also send to ACS for publication

July:

- Registration continues
- Contact Dr. Dervan and Dr. Searle personally over the phone

August:

- Send letters to students about the event and poster presentation
- Determine seminar professors to speak

September:

- Press release is released for the event again

Week of the Event
Monday:

- Send email confirming with graduate and undergraduate presenters

Tuesday:

- Confirm with Georgia Southern PAC and RAC for event preparation and catering
- E-mail reminding for volunteers

Wednesday:

- Confirm with seminar presenters
- Continue social media promotions
- Pick up banners for the event and set them up

Thursday:

- Confirm with venues and catering for food and timing
- Continue social media promotion

Friday: Kicking MBD1 to the Curb Event Begins

- 2:00 p.m. – Arrive at Georgia Southern RAC
- 2:30 p.m. – Speakers arrive
- 3:00 p.m. – Volunteers arrive
- 3:30 p.m. – Explain volunteers positions and run through the meeting
- 4:00 p.m. – Check-in begins; cookies and beverages are provided
- 6:00 p.m. – Seating for dinner and Seminars begin
- 6:15 p.m. – Welcome
- 6:30 p.m. – Dinner begins

- 7:00 p.m. – Brief overview of MBD1 and the great discovery followed by Dr. Mark searle talk and then open discussion
- 9:00 p.m. – Thank you's, goodnight's, and a brief discussion of itinerary for next day
- 9:15 p.m. – Event ends for the night
- 9:30 p.m. – Clean up

Saturday

- 8:00 a.m. – Graduate Poster Presentations and light breakfast
- 10:00 a.m. – Soccer tournament begins (Bracket found in appendix)
- 1:00 p.m. – Merchandise signed by Tormenta FC Players
- 2:00 p.m. – Signing ends
- 6:00 p.m. – Seminars provided by professors including Dr. Peter Dervan
- 7:00 p.m. – Standing for soccer tournament posted for following day
- 8:00 p.m. – Seminars by professors

Sunday:

- 8:00 a.m. – Undergraduate Poster Presentations and light breakfast
- 10:00 a.m. – Soccer tournament Continues (Bracket found in appendix)
- 2:00 p.m. – Championship game
- 3:00 p.m. – Champion and runner up awards
- 3:30 p.m. – Additional thank you, appreciation, and end event
- 4:00 p.m. – Clean up event space
- 5:00 p.m. - Thank volunteers
- 5:15 p.m. – Leave

Budget

Item	Cost
I. Promotional Material:	
Brochures/Flyers	Free (email)
Facebook Ads	$500.00
ACS Ads	$4,700
Event Invitations ($2.29*25)	$57.25
II. Event	
Venue ($1400*3days)	$4,200
Volunteers (food and T-shirts: $15 x 50)	$750
Grant Money for winning teams	$1000
Catering: Dinner ($12.95/pp*300)	$3885
Breakfast: ($7.25/pp*300ppl for 2 days)	$2175
T-Shirts for Sale (110 short and 70 long)	$2228.40
Fields (free to student/student organization 25live)	free
Concessions	$250
Total	$19,745.65

Reaching the Projected Budget

Item	Revenue
Registration Fee ($150*250people)	$37,500
Short Sleeve Shirts ($20-$10cost)*110	$1,100
Long Sleeve Shirts ($25-$15cost)	$700
Concessions	$500
Entry ($20*50people/day*2)	$2,000
Total	$41,800

Budgets do not include Banner sales to local businesses and groups. Banners will be sold as according to information in appendix, bringing in a revenue of $250-220 per banner

Conclusion

Evaluation

I. Impact Objective Evaluation:

a. Publicize the event in the ACS newsletter and journal, in collegiate campus newspapers around the world, and on Facebook in order to reach as many people as possible. Using big name biochemists such as Dr. Mark Searle and Dr. Peter Dervan will entice people to come and others will come just to play soccer. Either way, both tactics will make people want to come to the event.

b. The registration sheet asks questions if the individuals would be interested in playing soccer. Also, there are two grant prizes awarded to the first and second place team which should tempt people to play.

II. Output Objectives

a. Maintaining a consistent schedule will ensure that the status' are posted each week. Continuing to post the information about the event will help people to recognize Kicking MBD1 to the Curb and will increase interest. Multiple Facebook advertisements will be produced to prevent the same one from being shared every time.

b. Two press releases have been created in order to ensure at least two press releases are printed for MBD1 and the event. They will be sent to the Georgia Anne to be released around the World to captivate the readers and peak their interest of the event.

III. Informational Objective

a. The ads for the Georgia Anne are colorful and pleasing to the eye to attract attention. They will hopefully entice students to volunteer because of free food, needed volunteer hours, and free t-shirts. Posting multiple ads prior to the event will allow students taking summer classes (term A, B, and long term) and students enrolled in the fall of 2018 to have the opportunity to see the ads.

IV. Attitudinal Objectives:

a. In the registration form are questions asking the individual if they have gone to a conference before or if they plan to go to another. This information will be analyzed and then compared to the post-survey. It includes many of the same or similar questions. However, the question if the individual would want to attend another conference will be analyzed to see if the soccer event increases interest.

V. Behavioral Objectives:

a. The post-survey includes a question about how the individual perceived the conference as a whole. The goal is to ensure that at least 75% of the individual felt it at least exceeded their expectations. Having food at many of the seminars and poster presentations will limit the individuals from becoming hangry. Also, the soccer games will provide entertainment and excitement throughout the weekend, even for those that are not participating in playing.

Appendix

Event Documents
Logos

Kicking MBD1 to
the Curb
2687 Bunny Akins Boulevard,
Statesboro, GA 30458

Soccer Rules

6 vs. 6 Soccer Rules

All games will be governed by the 2018-2019 National Federation of State High School Associations Rules Book (NFHS) with the following Georgia Southern University Intramural Sports modifications:

Players & Equipment

1. Eligibility - See Intramural Handbook.

2. Each Men's and Women's team shall consist of 6 players each (including 1 goalkeeper).

Each team must have a minimum of 5 players in order to begin a game.

3. Due to injury, a team may continue with less than the minimum number of required players as long as that team has a chance to win. An ejection that leaves a team with less than the minimum number of required players will result in a forfeit by that team.

4. Each team is required to wear numbered shirts of one distinguishable color. Any team not dressed in like-colored shirts may wear the colored jerseys provided by Intramural Sports. Each goalie should wear a shirt that contrasts in color to that of the other players.

5. Shoes: Regulation, rubber-soled cleats, plastic cleats, detachable rubber cleats that screw into the shoe, and tennis shoes are the only permissible footwear. Sandals, street shoes, hiking boots, combat boots, or metal spikes are not allowed. No player will be allowed to participate in bare feet. No steel cleats or shoes with detachable steel cleats that screw onto the shoes may be worn.

6. Shin guards are recommended during play for personal safety.

7. Players may wear soft, pliable pads or braces on the leg, knee, and/or ankle. Braces made of any hard material must be covered with at least one-half inch of padding for safety reasons. Under no circumstances will a player wearing a cast or splint be permitted to play.

8. If eyeglasses are worn, they must be unbreakable. Each player is responsible for the safety of his/her own eyeglasses.

9. Jewelry: No jewelry or any other item deemed dangerous by the Intramural Staff may be worn. Any player wearing exposed permanent jewelry (i.e. body piercings) will not be permitted to play.

10. Headwear: Players may wear a knit or stocking cap (no caps with bills) during play. Bandanas which are tied with a knot are not permitted.

Game Format

1. The field will be modified to 80 x 40 yards.

2. Each game will consist of two 12-minute halves with a running clock. The clock will only stop for injuries. The game officials will be responsible for keeping the game clock.

3. Game time is forfeit time. A team must have the minimum number of players to start a game. If there is an insufficient number of players, the opposing captain has the option of taking the win or waiting for the minimum number of players to show. In the event that he/she decides to wait, that decision is irreversible and game clocks will be adjusted accordingly. If a team decides to wait, it will be required to wait a minimum of 12 minutes and play will start as soon as the opposing team has the minimum number of players present.

4. A coin toss at the beginning of the game shall determine which team has the choice of a goal to defend or kicking off first. The loser of the toss shall have the remaining option. Between halves, each team exchange ends and alternate the kickoff.

5. Mercy Rule: A game shall be called if a team is ahead by five (5) goals with five minutes or less remaining in the game and ten (10) goals with 7 minutes or less.

6. In the event that a score remains tied at the end of regulation play, the game will remain tied in regular season play. In a tournament game, a five minute sudden death extra period will be played followed by a shootout until a winner is determined. The shoot-out will proceed as follows:

A. The referee shall choose the goal at which all of the penalty kicks shall be taken.

B. Each captain will select any five different players on the field, including the goalkeeper, to take the penalty kicks.

C. The winner of a coin toss shall have the option of kicking first or second.

D. Teams will alternate kickers. There is no follow-up on the kick.

E. Following five kicks for each team, the team scoring on the greatest number of these kicks shall be declared the winner.

F. If the score remains tied after each team has had 5 penalty kicks, each team will select additional players (who were on the field at the end of regulation) to take kicks in a sudden death situation. Each team will alternate kicks until one team scores and the other team does not score, thus ending the game without more kicks being taken.

G. No player may take an additional kick until all those players who were listed on the scoresheet and present have kicked.

Start of Play

1. At the referee's signal, the game shall be started by a player taking a place kick into his/her opponent's half of the field of play (a forward pass). All players shall be in their team's half of the field and all players of the team opposing that of the kicker shall be at least 10 yards from the ball until it is kicked.

2. The kicker may not play the ball again on the kickoff until it has been touched or played by another player on either team. Penalty: Indirect free kick at the spot of the foul.

3. After a goal is scored, the team scored against shall restart play by a kickoff.

4. Between halves, teams will exchange goals to defend and the team who received first half will kick off second half.

Substitutions

1. Substitutions will be permitted after a score and at any goal kick. A team may also make a substitution during its own throw-ins and corner kicks and for an injured player. No substitutions may be made during a penalty kick. Players in the field may switch positions.

2. Substitutes must be recognized by the referee and must enter the field only after a player has left the field of play.

3. A substitute must enter for a player that has been cautioned (received a yellow card). The cautioned player may reenter the field of play at the next legal opportunity.

4. The goalkeeper may change positions with a player on the field during a stoppage of play or a substitution, provided the uniforms are legal and the official is notified prior to the change.

Scoring

1. A goal is scored when the entire ball passes legally beyond the goal line between the goal posts and under the cross bar, provided that it has not been carried, thrown, or propelled by the hand or arm. A ball on the goal line is not considered to have crossed the goal line.

2. If a defending player, other than the goalkeeper, intentionally stops the ball with his/her hands or arms to prevent a goal from scoring, then a goal is awarded and no penalty kick is awarded. Penalty: At the discretion of the official, the offending player may be presented with a red card and ejected from the game.

3. A goal MAY be scored during play directly from a:

    A. Kickoff

    B. Penalty Kick

    C. Corner Kick

    D. Drop Ball

4. A goal MAY NOT be scored during play directly from a:

    A. Indirect free kick

    B. Throw-in

    C. Free kick into a team's own goal

    D. Goal Kicks

Goalkeeper Play

1. The referee shall remove without caution any player who intentionally charges and contacts the goalkeeper. Warnings will be issued to players dangerously charging but not contacting the goalkeeper.

2. From the moment the goalkeeper takes control of the ball with the hands within his/her own penalty area, he/she has six seconds in which to release the ball into play. Possession includes holding, bouncing, or throwing the ball in the air and catching it again. After it has been released into play, the ball shall be played or touched by another player before the goalkeeper can touch it again with the hands.

3. The goalkeeper in possession of the ball must not be interfered with or impeded in any manner by an opponent.

4. On any occasion when a player deliberately kicks the ball to his/her own goalkeeper, the goalkeeper is not permitted to touch the ball with his/her hands. Penalty: Indirect free kick awarded to the opponent at the spot of the violation, unless in the goal area.

5. A goalkeeper shall not touch the ball with his/her hands when receiving it directly from a throw-in by a teammate.

6. The goalkeeper may not touch the ball with his/her hands or arms outside of the penalty box, but he/she may play the ball with any part of the body.

7. Goalkeepers may not punt the ball past midfield in the air.

Fouls and Misconducts

1. There will be no offside penalty.

2. A player shall be penalized if he/she:

    A. Handles the ball with his/her hands or arms. This does not apply to the goalkeeper within his/her own penalty area, provided he/she releases the ball within the prescribed six seconds.

    B. Trips an opponent, including throwing or attempting to throw an opponent by the use of the legs.

    C. Pushes or holds an opponent with the hand or with any part of the arm or body.

    D. Plays dangerously (kicks, strikes, attempts to kick or strike, jumps at an opponent, charges in a violent or dangerous manner, kicks dangerously high in front of opponent).

    E. Illegally obstructs an opponent by interfering with an opponent's movement without the ball.

3. A penalty shall also be assessed should two or more defensive players simultaneously make contact with the player who has control of the ball.

4. A player or coach will be cautioned (yellow card) for the following:

    A. Entering the field of play without the permission of an official.

    B. Persistent infringement of any of the rules of the game.

    C. Verbal objection or action indicating dissent toward the official.

    D. Unsportsmanlike conduct, which will be at the discretion of the official.

5. A player or coach will be ejected (red card) for the following:

    A. Exhibiting violent conduct or committing serious foul play including, but not limited to, deliberately handling a ball to prevent it from going into the goal or committing a foul against an opponent who is moving toward his/her goal with an obvious opportunity to score.

    B. Use of vulgar, profane, foul or abusive language.

    C. Fighting.

6. An ejected player cannot be replaced by a substitute and that team must play shorthanded.

Free Kicks

1. Free kicks shall be classified as: Indirect - two touches in which a goal cannot be scored unless the ball has been played or touched by a player other than the kicker before passing through the goal. All free kicks may be kicked in any direction from the point where the infraction occurred, except the penalty kick, which must be taken from the penalty spot and kicked forward. All free kicks are indirect. Free kicks are awarded for the following:

    A. Handling the ball with hands or arms.

    B. Tripping an opponent, including throwing or attempting to throw an opponent by the use of the legs and slide tackling.

    C. Pushing and opponent with the hand or with any part of the arm or body.

    D. Holding an opponent.

    E. Playing dangerously (kicks, strikes, attempts to kick or strike, jumps at an opponent, charges in a violent or dangerous manner).

    F. Charging an opponent in a dangerous manner.

    G. A player playing the ball a second time before it has been played by another player at the kickoff, free kick, a penalty kick, a corner kick, a goal kick, or by the thrower following a throw-in.

    H. Improper substitution.

    I. Persons other than authorized players entering the field.

    J. Dissension by word or action toward a referee's decision.

    K. Unsportsmanlike conduct.

    L. To resume play after a player is ordered off the field for persistent misconduct or violent conduct.

    M. Illegal obstruction (interfering with an opponent's movement without the ball).

    N. The goalie taking more than the allotted six seconds at any one possession.

    O. Charging the goalie or not allowing him /her to move with the ball.

2. When a free kick is being taken, a player of the opposite team shall not approach within 10 yards of the ball until it is in play. A violation of this may cause player removal from the game. The ball must be stationary when kicked, and after being kicked, the kicker shall not play the ball a second time until it has been touched by another player.

Penalty Kicks

1. A penalty kick shall be awarded when a foul, which ordinarily results in the awarding of a direct free kick, is committed by a defending player within his/her team's penalty area.

2. All players except the kicker and the opposing goalkeeper shall be within the field of play but outside the penalty area and at least 10 yards from and behind the penalty mark until the ball is kicked.

Throw-in

1. A throw-in shall be awarded when the opposing team last touches or plays the ball before the entire ball passes beyond the touchline either in the air or on the ground.

2. The ball shall be thrown in any direction from the point where it crossed the touchline by a player who is facing the field of play and has both feet on the ground on or behind the touchline. The thrower shall use both hands with equal force and deliver the ball from behind and over the head in one continuous movement.

3. On a throw-in, the ball is playable by either team when it has left the hands of the thrower and any part of it breaks the plane of the touchline.

Goal Kicks

1. A goal kick shall be awarded to the defending team when the entire ball crosses the goal line, excluding that area between the goal posts and under the crossbar, either in the air or on the ground, having last been played or touched by the attacking team.

2. Players opposing the kicker shall remain outside the penalty area until the ball has cleared the penalty area.

3. The ball shall be kicked from the ground from any point within the goal area by a player of the defending team. If the ball is not kicked beyond the penalty area, the goal kick shall be repeated.

Corner Kick

1. A corner kick shall be awarded to the attacking team when the entire ball passes over the goal line, excluding that area between the goal posts and under the crossbar, either in the air or on the ground, having last been touched or played by the defending team.

2. Players of the defending team shall be at least 10 yards from the ball until it has been kicked.

3. The ball shall be kicked from the ground within the quarter circle, including on the lines, nearest where the ball left the field of play.

4. After the corner kick, the ball may be played by any player except the one who executed the kick.

Stoppage of Play Due to Injury

1. Play will stop when the ball is out of play or at a time deemed suitable by the referee.

2. If the referee stops play, the game will re-start with a drop ball.

3. Injured players must stay on the field and on the ground.

4. Time will be stopped for medical attention, and play will resume as soon as the player is off the field.

Registration/Initial Survey

Kicking MBD1 to the Curb
2687 Bunny Akins Boulevard,
Statesboro, GA 30458

Name:
School:

Registration

Please take a moment to register for Kicking MBD1 to the Curb. It will help to cater the event to your needs and desires. Please attach your abstract to this document and email to KickingMBD@gmail.com. Mail the $150 registration fee along with an extra copy of this document to PO Box 2626, Statesboro Ga 30458.

Schooling and Research

How much schooling do you have?

- ☐ High School
- ☐ Some Undergraduate
- ☐ Undergraduate Degree
- ☐ Masters Degree
- ☐ PhD
- ☐ Professor

Are you interested in presenting?

- ☐ Yes
- ☐ No

If yes,

- ☐ Undergraduate Poster
- ☐ Graduate Poster
- ☐ Seminar

How many conference/seminars do you attend a year?

- ☐ This is my first
- ☐ 1-2
- ☐ 3-5
- ☐ More than 5

Do you think you will go to more conferences after this one?

- ☐ Never
- ☐ Not Likely
- ☐ I'm Not Sure
- ☐ Likely
- ☐ Of course

Soccer

Do you enjoy playing soccer?

- ☐ Yes
- ☐ No

Do you need a team for 6 versus 6 soccer?

- ☐ Yes
- ☐ No, more than 6 people are going from my school or research group

What is your research topic/concentration?

_____

_____

Do you have any food allergies or diet restrictions?

_____

About You (optional)

Name _____
Address _____
Email _____

May we add you to our mailing list, which offs news and exciting promotions?   ☐ Yes   ☐ No

Thank you for your participation!

Post-Survey

Kicking MBD1 to the Curb
2687 Bunny Akins Boulevard,
Statesboro, GA 30458

Follow Up Survey

Thank you so much for attending Kicking MBD1 to the Curb. Please take a moment to complete this survey.

Conference

How would you rate this conference?
- ☐ Exceptional
- ☐ Exceeds Expectations
- ☐ Meets Expectations
- ☐ Improvement Needed
- ☐ Unsatisfactory

Are you interested in the presentations?
- ☐ Yes
- ☐ No

If yes, which was your favorite
- ☐ Undergraduate Poster
- ☐ Graduate Poster
- ☐ Seminar

How many talks did you attend at the event?
- ☐ This is my first
- ☐ 1-2
- ☐ 3-5
- ☐ More than 5

Do you think you will go to more conferences after this one?
- ☐ Never
- ☐ Not Likely
- ☐ I'm Not Sure
- ☐ Likely
- ☐ Of course

Soccer

Did you enjoy playing soccer?
- ☐ Yes
- ☐ No

Do you need a team for 6 versus 6 soccer?
- ☐ Yes
- ☐ No, more than 6 people are going from my school or research group

Suggestions for improvement?

Food

Breakfast

Let us start your day the right way with one of our signature breakfasts. With these selections, you are sure to find just the right combination to satisfy any group at your event.

Continental

Our continental breakfast is a light and easy way to start your day. Served with fresh, seasonal fruit, orange juice, coffee (decaf on request) and your choice of the following items:

- Assorted Muffins

- Assorted Danishes

- Scones

Choose 1:	Choose 2:	Choose 3:
$7.25/person	$9.25/person	$11.25/person

Desserts

- Cheese Cake with Strawberry Topping

- Key Lime Pie

- Chocolate Cake

- Apple or Peach Cobbler

Italian Menu

All Italian Buffets come complete with Caesar Salad, Bread Sticks and your choice one Vegetable, and one Dessert.

Lasagna Buffet

Meat and Vegetable Lasagna

$12.95/person

Conference Hall Rental

Rental Price List

PLEASE NOTE: Per our pricing policy, prices listed are for reference only and are subject to change based on the number of event participants and the complexity of the event. This price list is not valid for University Academic or Departmental usage. The weekday rate applies to events on Monday-Thursday, and ending before 5:00 pm on Friday. The weekend rate applies to events on Friday after 5:00 pm, Saturday, and Sunday.

FACILITY FEES (Weekday M-Th):

ACTIVITY	FEE
Facility Rental (up to 8 hours)	$1,200.00
Facility Rental (per hour after 8 hours)	$150.00
Load In/Rehearsal (per hour up to 8 hours)	$100.00
Load In/Rehearsal (per hour after 8 hours)	$150.00

FACILITY FEES (Weekend F-Sun):

ACTIVITY	FEE
Facility Rental (up to 8 hours)	$1,400.00
Facility Rental (per hour after 8 hours)	$250.00
Load In/Rehearsal (per hour up to 8 hours)	$200.00
Load In/Rehearsal (per hour after 8 hours)	$250.00

LABOR:

STAFF	FEE
House Manager (per hour – 4 hour minimum)	$25.00
Ushers (per hour – 4 hour minimum)	$10.00
Technical Director (per hour after 8 hours)	$50.00
Box Office Manager (outside of regularly-scheduled hours)	$25.00
Additional Crew (when required)	Prevailing Wage

TECHNICAL SYSTEMS:

EQUIPMENT	FEE
Wireless Microphone (per unit)	$35.00
Use of Projector (per day)	$40.00
Use of Projector with Slide Advancer (per day)	$45.00

AA Batteries (per battery)	$2.00
9-Volt Batteries (per battery)	$3.00
Piano Tuning	$125.00
Color (per sheet)	$7.00
Gaffers Tape (per roll)	$18.00

Scripts

Welcome

The director will give opening remarks and welcome guests, colleagues and students

have I Feel Good by James Brown play to wake everyone up and prepare them for the welcoming speech.

Director begins

Welcome to Kicking MBD1 to the Curb an event that will change your lives. We have many exciting seminars set up for this weekend, along with some cutting edge research from undergrad and graduate students from around the world. Additionally, we have a new spin.

As many of you know the MBD1 protein binds methylated DNA and prevents the transcription of important proteins such as tumor suppressors. One of the reasons for this event tonight is to relish in the fact that Shelby Scherer of Georgia Southern University was able to publish an article in *JACS*. This research can be used as a stepping stone in the discovery of a protein that will bind methylated DNA at those CpG islands more effectively than the naturally occurs protein itself. Therefore, allowing the production of tumor suppressors.

Additionally, many of you may have heard we will be having a two day 6 versus 6 soccer tournaments. I would like to tell you there is a $750 research grant that will be presented to the winning team and a $250 research grant for the second place team. This is an exciting way to let out some endorphins, stay active, and stay focused at the seminars.

There are many speakers available and I would like to thank all of you for being here. I would now like to introduce our big named speaker, all the way from England, Dr. Mark Searle from the University of Nottingham.

Thanks, and Goodnight

Can we give another round of applause for our speakers tonight. Thank you all for coming and I am excited for this weekend. I think it is off to a great start. Games will begin tomorrow at 10AM after the graduate students present their posters at 8AM. There will be breakfast provided and then more seminars tomorrow night at 6pm.

Press Release

Journal Article Press Release

Fluorescence Makes MBD1 Shine

Developing a world without tumors is the long-time goal of a young scientist from Georgia Southern University. Shelby Scherer has spent years working with the protein known as MDB1, methylated CpG binding domain. She has finally determined a binding sequence that will lead to a peptide to kick the natural peptide out of the way. Preventing MBD1 from binding methylated DNA will prevent its function of stopping tumor suppressors.

MBD1 Protein Binding CpG Sites of Methylated DNA Exploration Using Florescence is the article that will be sweeping the nation, or at least the readers of the *Journal of the American Chemical Society*. The publication will be released in tomorrow, May 26th's issue. Ms.Scherer's research reasons through the heart of the binding sequence of the MBD1 protein and the structures binding of methylated DNA.

Ms. Scherer synthesized and purified the binding sequence of MBD1. The article discusses the use of cutting edge instrumentation such as a Fluorimeter and Circular Dichroism in order to visualize structure and binding. Fluorescence works similar to glow in the dark paint. A visible light is produced when the protein is not bound to the methylated DNA. The light strength decreases when the protein is bound to methylated DNA, providing efficient information. Circular Dichroism indicates the tertiary structure of the peptide which can be compared to the natural protein.

The research in this article provides a stepping stone to develop a more effective peptide as well as determine which amino acids in the protein are binding to the methylated DNA.

The use of the MDB1 protein is widely used to determine where methylated DNA is found. According to the study published in the February 13th issue of the *Journal of Neural Regeneration Research*, Dr. Agrawal, Dr. Ludwig, Dr. Thankam, Dr. Patil, and Dr. Chamczuk from the Department of Clinical and Translational Science at Creighton University School of Medicine indicate the necessity to clarify the underlying mechanism of the effect of MBD1. This portrays an additional example of how a new more effective peptide could be used. The new protein could take the place of MBD1 without having the negative effects of tumor production.

Event Press Release

Research Conferences Revolutionized

The research conference "Kicking MBD1 to the Curb" seems to be revolutionizing the way many people think of a scientific meeting. Kicking MBD1 to the curb is an event that is using a dynamic of combining the aspects of a normal conference with a soccer tournament. The conference is being held at Georgia Southern University's Performing Arts Center the weekend of October 12th-14th. Many well-known biochemists from around the world will be presenting their research such as Dr. Mark Searle from the University of Nottingham and Dr. Peter Dervan from California Institute of Technology.

The event aims to increase collaboration while simultaneously bringing scientists together to have fun and let loose playing soccer.

Kicking MBD1 to the Curb will be promoting the release of the article *MBD1 Protein Binding CpG Sites of Methylated DNA Exploration Using Florescence* by Shelby Scherer published in May 26ths issue of the *Journal of the American Chemical Society*. This journal article is the stepping stone for scientists attempting to mimic the MBD1 protein and study how it binds methylated DNA.

Fundraising

The 6 versus 6 soccer tournament is also a way to help fund research with a $750 grant for the first place team and a $250 research grant for the second place team. The games are to be played during the day on Saturday and Sunday.

At the event graduate posters will be presented Saturday at 8 a.m. and the undergraduate posters will be presented Sunday at 8 a.m.. The event is open to the public but there is a registration cost to participate and an entry fee of $20 per day at the door. The $150 registration fee is required to participate. The registration fee not only includes entry into the seminars and presentations throughout the weekend, but also includes dinner, two breakfasts, entry into the soccer tournament. In order to get more information about the event email kicking-MBD@gmail.com for a registration form or information.

Shirts

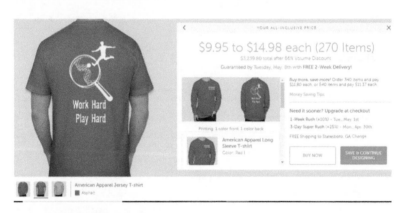

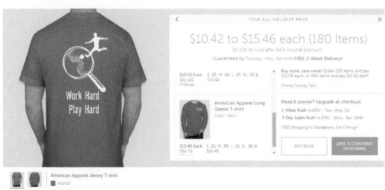

Banners

Banner Wholesale Cost

Quantity	Description	Unit Price	Total
9	Soccer Banners (Seniors)	40.00	360
7	Sponsorship Banners (full)	30	210
2	Half Banners	25	50
	.		
	Subtotal		
	Sales Tax		
	Shipping & Handling		
	Total Due		620.00

Due upon receipt

Thank you for your business!

Advertisement
Facebook
Facebook and ACS advertisements can be used interchangeably

Budget & Schedule
Define how much you'd like to spend, and when you'd like your ads to appear.

Budget ⓘ   Lifetime Budget ⬍ $500.00
                                 $500.00 USD

Schedule ⓘ   Start   📅 May 26, 2018   🕐 8:05 AM
             End   📅 Oct 12, 2018   🕐 8:05 AM
                     (Eastern Time)

Your ad will run until **Friday, October 12, 2018**.
You'll spend up to **$500.00** total.

Kicking MBD1 to the Curb

Bringing the love of science and soccer together for a weekend event of a lifetime. Seminars, posters, food and most importantly soccer are among some of the activities of the event occurring October 12-14th at Georgia Southern University.

Volunteer Ad

ACS Advertisement

UNIT RATES*

UNIT	1 PAGE	JUNIOR PAGE	1/2 PAGE	1/3 PAGE	1/4 PAGE
1X	$13,500	$9,375	$8,000	$5,750	$4,750
6X	$12,900	$8,925	$7,500	$5,250	$4,250
12X	$12,225	$8,250	$6,500	$4,750	$3,750
24X	$11,500	$7,500	$6,000	$4,250	$3,250

*Ad rates are per insertion in one publication and include 4-color process at no additional charge.

COLOR RATES

PMS color surcharge per unit: $1,820

INSERT RATES

| 2 PAGES | $6,460 per page | 6 PAGES | $4,120 per page |
| 4 PAGES | $5,290 per page | 8+ PAGES | Call for a Quote |

REPLY CARD INSERTS

| SINGLE | $5,050 | DOUBLE | $8,880 |

OUTSERTS & COVER TIPS

COVER RATES

All covers are: $14,025

Cover rates include space, position, bleed, and 1-color process or 4-color process.

SUPPLEMENTS: RIDE ALONG WITH C&EN ISSUES

FULL PAGE APP NOTE	$5,750
4-COLOR FULL PAGE AD	$7,350
2-PAGE APP NOTE	$8,750
FULL PAGE, 4-COLOR AD & FULL PAGE APP NOTE	$10,500

Event of the Year

Kicking MBD1 to the Curb is revolutionizing research conferences by including soccer tournament. Just us October 12th-14th at Georgia Southern University. Email KickingMBD@gmail.com for information

Subject Index